CLINICAL PROBLEMS IN STRABISMUS

Proceedings of an international symposium
identifying unsolved clinical problems
in strabismus
needing further research

The Smith-Kettlewell
Eye Research Institute

San Francisco

November 2012

Edited by Tina Rutar, M.D.

CLINICAL PROBLEMS IN STRABISMUS

Proceedings of an international symposium identifying unsolved clinical problems in strabismus needing further research

Held at The Smith-Kettlewell Eye Research Institute, San Francisco, California, November 7-9, 2012

Printed in the United States of America

ISBN: 978-0-9890819-2-4

Published by: The Smith-Kettlewell Eye Research Institute
                  2318 Fillmore Street
                  San Francisco, California 94115-1813

Cover design by: Gillian Gontard

TABLE OF CONTENTS

FOREWORD

This International Symposium arose from a desire to stimulate new research efforts in strabismus, by identifying key unsolved clinical problems that might be addressed or illuminated by scientific inquiry. The initial impetus, from discussions between Anja Palmowski-Wolfe, MD (Professor, University of Basel and former Institute fellow and staff researcher) and Arthur Jampolsky, MD (Institute Founder and current Board President), was translated into practice by Tina Rutar, MD (researcher on the Institute's Scientific Staff), who served as the Coordinating Organizer of the entire Symposium. An international advisory committee included, in addition to the above, Richard Harrad, MD, FRCS (Consultant Ophthalmologist at Bristol Eye Hospital and former Institute fellow and staff member) and John Brabyn, PhD (Institute Executive Director).

Suggestions for suitable unsolved clinical problems to be discussed were solicited from potential participants throughout the world, and collated by Tina Rutar, MD, into a coherent program for the two-day symposium. To maximize discussion, the round-table "workshop" format was utilized with presentations limited to 5 minutes each. The full 200-page proceedings from these round table discussions and the statements and opinions of the 45 participating strabismus experts from around the world were transcribed by Elizabeth Warner and condensed into the present volume by Dr. Rutar. The full transcription will be made available online.

I would like to take this opportunity to thank everyone involved in the Symposium, including the many participants who traveled long distances and the numerous staff who assisted with arrangements and logistics. Above all, one individual towers above the rest; without the overall coordination, energy and efforts of Dr. Tina Rutar the enterprise would not have succeeded. We are fortunate indeed that she undertook this very considerable task, and are confident that the final result, in the form of the present volume, will serve as a landmark in the field and help provide inspiration for both basic and clinical researchers to study some of the important and fascinating problems in strabismus that still elude solutions.

John Brabyn, PhD
Executive Director
The Smith-Kettlewell Eye Research Institute
San Francisco

SYMPOSIUM PARTICIPANTS

Harley E. A. Bicas, M.D.
Full Professor of Ophthalmology
Department of Ophthalmology
Medical School of Ribeirao Preto
University of São Paulo
São Paulo, Brazil
heabicas@fmrp.usp.br

Bradley C. Black, M.D.
Clinical Assistant Professor
Department of Ophthalmology
Louisiana State Univ. Health Sciences Center
The Pediatric Eye Care Center
Baton Rouge, Louisiana
bcblack2000@me.com

John A. Brabyn, Ph.D.
Executive Director
Senior Scientist
Smith-Kettlewell Eye Research Institute
San Francisco, California
brabyn@ski.org

Michael C. Brodsky, M.D.
Professor of Ophthalmology and Neurology
The Mayo Clinic
Rochester, Minnesota
brodsky.michael@mayo.edu

Arvind Chandna, M.D., D.O., F.R.C.S.
Consultant Ophthalmologist
Pediatric Ophthalmology and Adult Strabismus
Alder Hey Children's Hospital
Royal Liverpool & Broadgreen University Hospital
Liverpool, United Kingdom
Arvind.Chandna@alderhey.nhs.uk

Alberto O. Ciancia, M.D.
Past President
Consejo Latino-Americano de Estrabismo
International Strabismological Association
Buenos Aires, Argentina
a.ciancia@hotmail.com

Michael Clarke, M.B., BChir, FRCOphth
Consultant Paediatric Ophthalmologist
Newcastle upon Tyne Hospitals
NHS Foundation
Trust/Reader in Ohthalmology
Newcastle University
Newcastle upon Tyne, UK
Michael.Clarke@newcastle.ac.uk

Alejandra de Alba Campomanes, M.D., M.P.H.
Assistant Professor of Ophthalmology
University of California, San Francisco
San Francisco, California
dealbaa@vision.ucsf.edu

R. Scott Foster, M.D.
Clinical Professor of Ophthalmology
Stanford University
Stockton, California
rsf@cvemg.com

Douglas R. Fredrick, M.D.
Clinical Professor of Ophthalmology
Stanford Hospital and Clinics
Byers Eye Institute
Stanford University Medical Center
Stanford, California
dfred@stanford.edu

Mauro Goldchmit, M.D.
Professor of Ophthalmology
Santa Casa de São Paulo Hospital
and Cema Institute
São Paulo, Brazil
maurog@uol.com.br

William V. Good, M.D.
Senior Scientist
Smith-Kettlewell Eye Research Institute
San Francisco, California
good@ski.org

Michael Graef, M.D.
Professor
Justus-Liebig University Giessen
Giessen, Germany
Michael.H.Graef@augen.med.uni-giessen.de

David L. Guyton, M.D.
Zanvyl Krieger Professor of Ophthalmology
The Krieger Children's Eye Center
at The Wilmer Institute
Baltimore, Maryland
dguyton@jhmi.edu

Gunilla Haegerstrom-Portnoy, O.D., Ph.D.
Senior Scientist
Smith-Kettlewell Eye Research Institute
Professor, School of Optometry
University of California, Berkeley, California
ghp@berkeley.edu

Richard Harrad, M.D., F.R.C.S.
Consultant Ophthalmologist
Bristol Eye Hospital
Bristol, United Kingdom
R.A.Harrad@bristol.ac.uk

Jonathan C. Horton, M.D., Ph.D.
William F. Hoyt Professor of Ophthalmology
University of California, San Francisco
San Francisco, California
hortonj@vision.ucsf.edu

Arthur Jampolsky, M.D.
Founder
President, Board of Directors
Smith-Kettlewell Eye Research Institute
San Francisco, California
aj@ski.org

Robert J. Johnson, M.D.
Ophthalmologist (retired)
Sausalito, California
rjmd2020@comcast.net

Stephen P. Kraft, M.D.
Professor of Ophthalmology
University of Toronto
Toronto, Canada
stephen.kraft@sickkids.ca

K-Min Lee, M.D., Ph.D.
Professor of Neurology
Seoul National University
Seoul, South Korea
kminlee@snu.ac.kr

Lora T. Likova, Ph.D.
Associate Scientist
Smith-Kettlewell Eye Research Institute
San Francisco, California
lora@ski.org

Suzanne P. McKee, Ph.D.
Senior Scientist
Smith-Kettlewell Eye Research Institute
San Francisco, California
suzanne@ski.org

Keith W. McNeer, M.D.
Clinical Professor of Ophthalmology
Virginia Commonwealth University
Medical College of Virginia
Richmond, Virginia

Henry S. Metz, M.D., M.B.A.
Executive Director, Emeritus
Smith-Kettlewell Eye Research Institute
San Francisco, California
henry@ski.org

Hermann Mühlendyck, M.D.
Professor of Ophthalmology, Emeritus
Medical School of Göttingen
Göttingen, Germany
hermann-muehlendyck@kabelmail.de

Anthony (Tony) M. Norcia, Ph.D.
Professor (Research)
Department of Psychology
Stanford University
Stanford, California
amn@ski.org

Omondi Nyong'o, M.D.
Pediatric Ophthalmologist
Palo Alto Foundation Medical Group
Palo Alto, California
kideyedoc@gmail.com

J. Vernon Odom, Ph.D.
Professor of Ophthalmology,
Physiology, and Pharmacology
West Virginia University
Morgantown, West Virginia
odomj@wvuhealthcare.com

Anja Palmowski-Wolfe, M.D.
Professor Doctor
University of Basel
Basel, Switzerland
apalmowski@uhbs.ch

Cameron F. Parsa, M.D.
Associate Professor of Ophthalmology
University of Wisconsin-Madison
parsa@wisc.edu

David A. Romero Apis, M.D.
Past President
Consejo LatinoAmericano de Estrabismo
Mexican Ophthalmological Society
Queretaro City, Mexico
descarra@prodigy.net.mx

André Roth, M.D.
Hon. Professor of Ophthalmology
Medical School of Geneva
Geneva, Switzerland
andrech.roth@gmail.com

Tina Rutar, M.D.
Scientific Staff Researcher
Smith-Kettlewell Eye Research Institute
San Francisco, California
tinarutar@yahoo.com

Denise Satterfield, M.D.
Associate Clinical Professor
UC Davis Eye Center
University of California, Davis
Sacramento, California
docdenise@hotmail.com

Alan B. Scott, M.D.
Senior Scientist
Smith-Kettlewell Eye Research Institute
San Francisco, California
abs@ski.org

Felisa Shokida, M.D.
Consultant, Pediatric Ophthalmology
and Strabismus
Ophthalmology Service
Hospital Italiano de Buenos Aires
Buenos Aires, Argentina
felisa.shokida@gmail.com

Huibert J. Simonsz, M.D., Ph.D.
Professor
Department of Ophthalmology
Erasmus Medical Center
Rotterdam, Netherlands
simonsz@compuserve.com

John Sloper, FRCOphth
Consultant Ophthalmic Surgeon
Moorfields Eye Hospital
London, United Kingdom
John.Sloper@moorfields.nhs.uk

Carlos R. de Souza-Dias, M.D.
Professor, Faculty of Medical Sciences
of the Santa Casa of São Paulo
São Paulo, Brazil
csdias@uol.com.br

Christopher W. Tyler, Ph.D., D.Sc.
Senior Scientist
Smith-Kettlewell Eye Research Institute
San Francisco, California
cwt@ski.org

Federico G. Vélez, M.D.
Asst. Clinical Professor of Ophthalmology
Jules Stein Eye Institute
University of California, Los Angeles
Los Angeles, California
velez@jsei.ucla.edu

Gerald Westheimer, Ph.D.
Professor of the Graduate School
            Division of Neurobiology
Department of Molecular & Cell Biology
Clinical Professor, School of Optometry
University of California, Berkeley
Berkeley, California
gwestheimer@berkeley.edu

Maria Arroyo Yllanes, M.D.
Chief of Ophthalmology Department
Hospital General de México
Mexico City, Mexico
mearroyo1@gmail.com

SMITH KETTLEWELL EYE RESEARCH INSTITUTE
CLINICAL PROBLEMS IN STRABISMUS
DETAILED SCIENTIFIC PROGRAM

WEDNESDAY, NOVEMBER 7ᵀᴴ   WELCOME

9:00 am – 9:15 am   Opening remarks

WEDNESDAY, NOVEMBER 7ᵀᴴ   ESOTROPIAS
(MODERATORS: TINA RUTAR, DOUG FREDRICK)

9:15 am - 10:30 am   Infantile esotropia

John Sloper: How should we optimally manage infantile esotropia? Should these babies be treated with surgery or botulinum toxin injection?

Why this is a problem: Both surgery and botulinum toxin require general anesthesia. Patients require reoperations for consecutive exotropias, DVD, and V patterns. Patients may require multiple botulinum toxin injections, and botulinum injection can be complicated by ptosis.

Alberto Ciancia: Infantile esotropia and late onset esotropia.

André Roth: What are the best motor and sensory outcomes?

Michael Clarke: Which clinical factors promote stability of ocular alignment?

Huibert Simonsz: Individual timing of surgery for infantile esotropia according to treatment targets.

Mauro Goldchmit: What is the best timing of surgery for infantile esotropia?

Keith McNeer: Advantages of botulinum toxin.

1

Richard Harrad: How can we get good surgical results without creating consecutive esotropias?

Additional comments: Alan Scott, Alejandra de Alba Campomanes, Harley Bicas

10:30 am – 10:45 am   **Break**

10:45 am – 11:30 am   **Infantile esotropia**

Moderator: What is the underlying cause of infantile esotropia in the neurologically healthy child without significant hyperopia and without differences in input between the eyes? Could better understanding of the cause of infantile esotropia be used to improve treatment methods?

Why this is a problem: We do not tailor treatment to the underlying cause.

Michael Brodsky: Does infantile esotropia have a cortical or subcortical origin?

Arthur Jampolsky: The role of increased medial rectus muscle tone. Why does this occur?

Michael Brodsky: Why do dissociated vertical deviation (DVD), latent nystagmus and inferior oblique overaction frequently coexist with infantile esotropia?

Additional comments: Gerald Westheimer, Cameron Parsa

11:30 am – 11:45 am   **Break**

11:45 am – 12:30 pm   **Accommodative esotropia**

Moderator: How can we optimize the correction of hyperopia in the treatment of accommodative esotropia?

Why this is a problem: Not all strabismologists agree that accommodative esotropia should be treated with full plus refractive correction. Some children do not tolerate full plus refractive correction. Accurate retinoscopy in young children is difficult and depends on the skill of the examiner.

Scott Foster: Inadequate assessment of the full hyperopic refractive error and inadequate duration of refractive correction lead to unnecessary surgeries and esotropia recurrences.

Bradley Black: Undercorrection of hyperopia does not promote emmetropization.

Tina Rutar: How can one relax accommodation in patients who habitually accommodate? Cycloplegic agents often don't work for older children and adults.

Additional comments: Hermann Mühlendyck, Huibert Simonsz, Carlos Souza-Dias

12:30 pm – 1:00 pm   Esotropia at near

Moderator: What is the best management strategy for esotropia at near (high accommodative convergence/accommodation ratio esotropia) in children who are already wearing their full hyperopic refraction, or who have negligible hyperopia?

Why this is a problem: Young children may not wear bifocal spectacles or use them appropriately. Phospholine iodide has side effects and is difficult to obtain. Does surgery appropriately eliminate the excess esotropia at near?

Hermann Mühlendyck: Can the paretic effect of the Faden operation be better quantified by the reduction of the arc of contact or by the Demer procedure?

3

Arvind Chandna: Loop recession for the management of near esotropia.

Additional comments: Anja Palmowski-Wolfe

WEDNESDAY, NOVEMBER 7TH   BINOCULAR   VISION, AMBLYOPIA, AND SUPPRESSION   (MODERATORS: RICHARD HARRAD, BRADLEY BLACK)

2:00 pm – 2:45 pm   Amblyopia assessment

Arvind Chandna: Can amblyopia be better assessed and monitored by using approaches other than measurement of high contrast visual acuity?

Why this is a problem: Most clinics are not equipped to measure contrast sensitivity, grating acuity, Vernier acuity, visual evoked potentials, etc. Subtypes of amblyopia differ in their effects on these parameters. Visual acuity testing is unreliable in young children.

William Good: Will VEP testing become more mainstream, performed quickly during routine clinic visits?

Additional comments: Suzanne McKee, K-Min Lee, Vernon Odom, Gunilla Haegerstrom-Portnoy, Michael Graef

2:45 pm – 3:30 pm   Amblyopia treatment and **cortical plasticity**

Moderator: Amblyopia treatment is primarily limited to refractive correction, occlusion, penalization, and correction of strabismus. Treatment is most effective when instituted at a young age. How can we improve upon these limited treatment options?

Why this is a problem: Amblyopia may improve but not resolve, especially with late diagnosis. Treatment compliance is poor.

Cameron Parsa: How are issues of inhibition and competition addressed in new treatments, such as acupuncture and systemic levodopa? Do these treatments work?

Anja Palmowski-Wolfe: How might plasticity of the visual cortex be increased beyond childhood years?

Felisa Shokida: Can strabismus surgery improve binocular cortical activation in adult patients?

Additional comments: Stephen Kraft, Arvind Chandna, Lora Likova

3:30 pm – 3:45 pm   Break

3:45 pm – 4:45 pm   Suppression

Moderator: Suppression is friend and foe. Amblyopia treatment often involves breaking down suppression that has developed as a compensatory mechanism to avoid diplopia and visual confusion.

Why this is a problem: We could use a better understanding of suppression to our advantage in treating amblyopia.

Richard Harrad: What is the physiological basis for suppression in strabismus?

Richard Harrad: What is the role of suppression in the pathogenesis of amblyopia?

Stephen Kraft: Why does suppression develop to different depths in different patients?

Arvind Chandna: How do you measure suppression? Can you use depth of suppression to predict improvement in monocular visual acuity and binocular function in anisometropic amblyopia, for example?

John Sloper: Is it necessary to break down suppression to treat amblyopia in adults with strabismus?

Additional comments: Jonathan Horton, Tony Norcia, Suzanne McKee

4:45 pm – 5:45 pm   Diplopia

Richard Harrad: When treating adults with strabismus, we cannot accurately predict who is at risk for postoperative diplopia, nor its severity.

Why this is a problem: The fear of postoperative diplopia leaves some adults with uncorrected strabismus, and thus they do not experience optimal binocular function nor the psychosocial benefit of straight eyes.

Moderator: How can we better manage intractable diplopia in adult patients? Can adults be taught to suppress?

Denise Satterfield: Practical management strategies. Monovision as an alternative to occlusion.

Additional comments: David Guyton, John Sloper

Doug Fredrick: How can we better prevent or manage the fixation switching diplopia that some patients experience?

THURSDAY, NOVEMBER 8TH   INTERMITTENT EXOTROPIA (MODERATORS: ANJA PALMOWSKI-WOLFE, BILL GOOD)

8:30 am – 9:30 am   Intermittent exotropia

Moderator: Surgical results in intermittent exotropia are unstable. How can we better determine the timing and the type of intervention?

Why this is a problem: Exotropia frequently recurs, and

patients require reoperations. Overcorrections in patients who are young can cause suppression and development of amblyopia, and overcorrections in adults result in diplopia.

Scott Foster: How can you predict preoperatively who is likely to recur?

Anja Palmowski-Wolfe: Does early surgery lead to better outcomes?

Michael Clarke: How should a randomized controlled trial be designed to assess whether early surgical treatment of intermittent exotropia leads to better outcomes?

John Sloper: Should we overcorrect intermittent exotropes and by how much?

Anja Palmowski-Wolfe: In those who are overcorrected, should we let fusion drive realignment, or provide prisms to prevent suppression?

Cameron Parsa: Should we deliberately make young children monofixators?

Additional comments: Federico Vélez

9:30 am – 10:15 am   Intermittent exotropia

Moderator: Inadequate understanding of the underlying cause of intermittent exotropia may be handicapping our treatment methods and success.

Jonathan Horton: Suppression in intermittent exotropia. Does understanding suppression provide any clues to the underlying cause?

Michael Brodsky: The role of luminance on eye position in normal patients. Might the retinal ganglion cell pathway be implicated in intermittent exotropia?

David Romero Apis: How are intermittent exotropia and dissociated horizontal deviation different? Does that tell us anything about underlying cause?

Christopher Tyler: Intermittent exotropia: can fMRI be used to study the underlying mechanism?

10:15 am – 10:30 am   Break

THURSDAY, NOVEMBER 8ᵀᴴ   OBLIQUE   DYSFUNCTIONS (MODERATORS: ANJA PALMOWSKI-WOLFE, HENRY METZ)

10:30 am – 11:45 am   Oblique overactions, A and V patterns

Michael Clarke: Under what circumstances should inferior oblique surgery be performed when inferior oblique overactions coexist with childhood strabismus?

Why this is a problem: Surgical correction increases anesthesia time, and adds some additional operative risks, whereas lack of surgical correction may require a second surgery in the future.

Maria Arroyo Yllanes: Clinical tests for oblique overaction and surgical indications.

Additional comments: Omondi Nyong'o, Scott Foster, Maria de Alba Campomanes

Moderator: What causes A and V patterns?   Could knowledge of the cause allow us to better predict who will develop A and V patterns, and how they should optimally be treated?

Why this is a problem: Patients who develop these patterns

may require additional surgeries. Patients who fuse and have A or V patterns may adopt anomalous head postures.

André Roth: What is the appropriate surgical strategy for oblique overactions coexisting with horizontal deviations?

Henry Metz: In A and V patterns, why does a vertical shift of the horizontal recuts muscles eliminate or minimize the apparent oblique overaction?

1:30 pm – 1:45 pm   **Group photo**

1:45 pm – 3:00 pm   **Superior   oblique   palsy, Brown's syndrome**

David Guyton: What causes congenital "superior oblique palsy," and should one tailor treatment to the underlying cause?

Why this is a problem: Not knowing the cause could lead to inappropriate/incorrect surgery.

Felisa Shokida: Is the contralateral superior oblique normal in the patient with congenital superior oblique palsy?

Additional comments: Cameron Parsa, Mauro Goldchmit, Huibert Simonsz

Hermann Mühlendyck: What is the pathogenesis underlying congenital Brown's syndrome?

Why this is a problem: Not knowing the cause could lead to inappropriate/incorrect surgery.

Additional comments: Michael Graef

Moderator: How should one treat bilateral superior oblique palsy?

Why this is a problem: Various surgical options may not address both the excyclotorsion and the V pattern esotropia, leaving patients diplopic in primary gaze and/or down gaze.

Anja Palmowski-Wolfe: Harada-Ito procedure and other surgical strategies.

Carlos Souza-Dias: Surgical management with large bilateral inferior rectus recessions.

3:00 pm – 3:15 pm   **Break**

THURSDAY, NOVEMBER 8TH   DVD   (MODERATORS: TINA RUTAR, HENRY METZ)

3:15 pm – 4:00 pm   Dissociated   Vertical Deviation (DVD)

Doug Fredrick: Can we come up with a better solution for DVD other than surgically limiting ocular rotations?

Why this is a problem: DVD can recur unless you cripple the ability of the eye to move up, or establish sensory fusion. Limiting ocular rotations is not an ideal solution to the problem. Many patients with DVD do not have the ability to fuse.

Maria Arroyo Yllanes: Clinical tests in the diagnosis of DVD and indications for surgical treatment.

David Guyton: Can we treat the underlying cause of DVD?

David Romero Apis: Innervational surgical approach for DVD in congenital esotropia.

THURSDAY, NOVEMBER 8TH   NEW   APPROACHES   TO STRABISMUS   MANAGEMENT   (MODERATORS: JOHN BRABYN, MARIA ARROYO YLLANES)

4:00 pm – 5:00 pm   New   approaches   to **strabismus management**

Moderator: Strabismus management is essentially limited to refractive correction, amblyopia treatment, extraocular muscle injections, and surgery. How can we develop new solutions?

Why this is a problem: Other fields in ophthalmology have had treatment revolutions, but we are still employing many techniques developed decades ago.

Harley Bicas: How to eliminate unwanted eye movements, including postoperative drift.

Cameron Parsa: How can we determine factors that affect innervational muscle tonus and modulate these clinically in the treatment of strabismus?

André Roth: Can the variance in surgical results be reduced by taking into account intraoperative data?

David Guyton: How can we utilize muscle length adaptation in the prevention and treatment of strabismus?

Federico Vélez: An electronic stimulator implantable into extraocular muscles for the treatment of muscle palsies.

Scott Foster/Alan Scott: A spring-loaded device for the treatment of duction deficits.

ESOTROPIAS

INFANTILE ESOTROPIA

1. **Clinical Problem: How should we optimally manage infantile esotropia? Should these babies be treated with surgery or botulinum toxin injection?**

Why this is a problem: Both surgery and botulinum toxin require general anesthesia. Patients require reoperations for consecutive exotropias, dissociated vertical deviations (DVD), and V patterns. Patients may require multiple botulinum toxin injections, and botulinum injection can be complicated by ptosis.

John Sloper said that the logic behind early surgery in infantile esotropia is achieving a better binocular sensory outcome. To this regard, he reviewed Hubel and Wiesel's data on the development of ocular dominance columns after temporary closure of one eyelid. The sensitivity period for developing binocularly driven cells in the visual cortex differs between monkeys and humans: the sensitive period in monkeys is approximately up to the age of three months, but it is approximately between four and 18 months in humans. If you get the eyes straight enough, early enough, can you get stereopsis? The means by which you achieve this, whether surgery or botulinum toxin, are less important than the timing. When do you know that a baby has permanent strabismus? Should you wait until the angle has stabilized and does doing so make you delay intervention past the sensitive period for developing optimal binocular vision?

Alberto Ciancia defined the clinical features of infantile esotropia, which he believes is a clinically unique form of strabismus as:

a. Onset within the first six months of life.
b. Large angle of deviation.
c. The head is turned towards the side of the fixing eye

13

and often tilted towards the same side.

d. Jerk nystagmus with the rapid phase towards the side of the fixing eye that increases in abduction and decreases in adduction.

e. Rotatory nystagmus, with intorsion of the fixing eye and extorsion of the non-fixing eye.

f. Dissociated vertical, torsional, and horizontal deviations (DHD).

g. Latent nystagmus.

h. A particular pattern of eye movement oculography that differs markedly from the eye movement oculography in late onset esotropia cases.

i. A visual evoked potential (VEP) that differs from late onset esotropia cases (asymmetry).

j. Ophthalmoscopic micro nystagmus.

k. Delayed maturation of visual acuity in both eyes.

André Roth discussed the difference between postoperative true orthotropia, which allows for development of normal binocularity, versus microtropia, and development of abnormal binocularity.

Michael Clarke asked which clinical factors promote stability of ocular alignment? Fusional vergences maintain ocular alignment in visually mature individuals, but to what extent is that true in the visually immature child presenting with infantile esotropia? Predictors of stability in ocular alignment include good surgical outcome, correction of refractive error, and presence of fusion.

Huibert Simonsz stated that infantile esotropia is not one disease, but a collection of disorders, and one must tailor the treatment to the individual patient. Here is a problem of semantics: should infants with esotropia presenting within the first six months with high hyperopia be included in our discussion of infantile esotropia? What about esotropia presenting within the first six months in the neurologically impaired child? What about transient esotropia presenting within the first six months? What about a small angle or an intermittent esotropia? These don't share the clinical features of infantile esotropia outlined by Alberto Ciancia.

Dr. Simonsz reviewed the data from the European Early vs. Late Infantile Esotropia Surgery Study. Early surgery (operation at mean age of 20 months) compared to late surgery (operation at mean age of 49 months) is associated with better stereopsis but higher reoperation rates. With early surgery, there is a risk of operating on a patient who may spontaneously resolve.

Mauro Goldchmit presented data on relatively poor motor outcomes on patients operated on for infantile esotropia: the longer you follow them, the worse they do. He also reviewed the literature of Ken Wright, Malcolm Ing, Julio Prieto-Díaz, and Agnes Wong regarding the timing of surgery and the potential for binocular vision. A review of the literature did not provide a clear consensus regarding optimal timing for development of binocular vision. Practically, Dr. Goldchmit finds that he rarely gets a child to the operating room before one year of age, in part because he required fixation to freely alternate and to have reliable, stable measures of esotropia, and these do not occur until seven to eight months of age.

Keith McNeer presented his twenty-year experience in treating patients with infantile esotropia with botulinum toxin injection. First, he institutes alternate full day occlusion to "erase anomalous pathways." He then injects 2.5 units of botulinum into each medial rectus muscle under brief general anesthesia with nitrous oxide. He believes it is important to create fixation duress with a consecutive exotropia. With either eye fixing, the other eye will exhibit a large angle exotropic. His outcomes include stereopsis of >400 arc seconds in 34% of patients and 40 arc seconds in 40% of patients. He saw a 2% rate of permanent consecutive exotropias.

Discussion:

Arthur Jampolsky remarked on the importance of alternate full-day patching of babies until one can operate: this practice prevents abnormal visual input/development.

Jonathan Horton stated that data in primates do not support alternate occlusion, and alternate occlusion can induce strabismus; some become exotropic, esotropic, or variable types of strabismus. The benefit of alternate patching is that if you don't know whether amblyopia is present, it amounts to prophylactic amblyopia treatment. However, it promotes breakdown of the binocular connections in the cortex.

Scott Foster said that these kids already have strabismus.

Jonathan Horton confirmed that, yes, these kids already have strabismus, but strabismus in some children might spontaneously improve after four to six months, and alternate patching would be disruptive to development of binocularity in those children.

Stephen Kraft stated it's more important to look at the age when you achieve orthotropia rather than the age at first surgery. If you do the first surgery near the end of the window of opportunity, you may not have time for a reoperation, if needed, before that window is closed.

Keith McNeer said that timing is critical because that window closes at approximately eight months of age. Alternate patching is of real value in stopping anomalous central nervous system pathways.

David Guyton said that we should not assume that these infants have a normal fusional vergence system.

John Sloper said a defective fusional vergence system causes variability in the angle of early strabismus.

David Romero Apis remarked that the surgical result must be perfect orthotropia, which is not achievable.

Jonathan Horton said that three common clinical problems he encounters in reviewing the medical records of patients treated for infantile esotropia: (a) after surgery, the surgeon measures the postoperative deviation far less accurately than the preoperative deviation and does not correct small residual deviations (e.g. with prisms), (b) usually, the methods to test stereopsis are rudimentary and not accurate, and (c) we rely heavily on the parental perception of a good outcome due to improvement in appearance of alignment.

André Roth said that it's important to first correct the ametropia and maintain full optical correction. Second, the patient should be freely alternating. He prefers to operate after age two years, when the basic angle of deviation is known.

Mauro Goldchmit stated that in terms of binocular alignment, complexity in the type of strabismus, nystagmus, DVD, etc., makes it difficult to obtain binocularity.

Alberto Ciancia said that Panum's area is no more than ½ PD wide. He has operated on over 1,000 infantile esotropia patients and never had a normal binocular outcome.

Keith McNeer said that if you use Botox, you can achieve binocularity and avoid development of late sequelae such as DVD, V pattern with inferior oblique overaction, and latent nystagmus.

John Sloper remarked that one of the benefits of Botox is that you make them exotropic transiently and the patient will achieve true orthotropia at some point as the eyes drift back in.

Huibert Simonsz stated that the central nervous system may not be normal in these patients.

Stephen Kraft said that 30 years ago, strabismologists waited for stability before intervening. Pediatric Eye Disease Investigator Group (PEDIG) studies have shown that when you carefully measure the angle in infantile esotropes, it varies by 10 PD from visit to visit in a significant minority of patients. "Variability of angle" should be added to the clinical characteristics of infantile esotropia.

Keith McNeer remarked that Botox is self-adjusting.

Alan Scott said that one of the benefits of Botox is that you could treat the baby at an early age, before the angle becomes stable. With surgery, it's preferable to wait to have a stable angle and to confirm that the child is normal neurologically. Botox can be done at three months of age and has a good safety profile. Keith McNeer showed that he has only a 2% rate of permanent consecutive exotropia, which is much better than with surgery. Alan Scott reports that he can perform Botox injections into the medial rectus muscles in babies under the age of six months in the office without anesthesia.

Babies with esotropia who are three to five months of age have the potential to develop sensory fusion, and thus Botox is the best treatment for them. In babies who have no potential for sensory fusion, he prefers surgery.

Alejandra de Alba Campomanes agreed that Botox injected early "does no harm," even if the child is later found to be neurologically abnormal. She uses five units into each medial rectus muscle. In her published study from Mexico, she noted a 3% rate of permanent consecutive exotropias.

2. Clinical Problem: What is the underlying cause of infantile esotropia in the neurologically healthy child without significant hyperopia and without differences in input between the eyes? Could better understanding of the cause of infantile esotropia be used to improve treatment methods?

Why this is a problem: We do not tailor treatment to the underlying cause.

Michael Brodsky described that in early infancy, dissociated binocular vision alters the balance of excitation and inhibition to both cortical and subcortical visual pathways, allowing cortical suppression to potentiate the activity of subcortical visual input, which is normally functional in the first two to three months of human life.

DVD, primary inferior oblique muscle overaction, and latent nystagmus all correspond to subcortical (visuo-vestibular) reflexes that are operative in lateral-eyed animals. DVD occurs in the roll plane, primary oblique muscle overaction (POMA) occurs in the pitch plane, and latent nystagmus in the yaw plane. Any combination of dissociated eye movements can be represented vectorially as a single central vestibular imbalance in a 3-dimensional space. DVD, POMA, and latent nystagmus all arise from the same binocular visual imbalance in different planes.

Arthur Jampolsky remarked that increased medial rectus tonus may be at play in some infantile esotropia (with zero refractive error) who have some form of cerebral palsy and hyperactivity. This excessive esotropia is non-accommodative and nonretinal. An infant with a 50-60 PD of infantile esotropia may become exotropic in the operating room under anesthesia. Amazingly, the entire significant degree of esotropia completely disappears under anesthesia in these specific patients. At surgery, the surgeon may elect to make this patient exotropic on the operating table (no refractive error) so that "hopefully" the medial rectus esotonus will appropriately realign the eyes postoperatively. Such is the abysmal state of strabismus management in this very selected patient. The same surgical approach in an adult esotropic patient, who does not have the same medial rectus hyper-esotonus and hyperactivity when awake, would not work. A robust discussion on the role of muscle tonus can be found on the Smith-Kettlewell website http://www.ski.org/Tonus/index.html. This type of hypertonus (non-accommodative) in usually hyperactive patients should not be considered in that millions of child esotropes have cortical damage (in the absence of demonstrable findings).

Discussion:

Stephen Kraft remarked that what impresses him about Dr. Brodsky's description is that the disorders seen in humans are the result of residual pathways and evolutionary development from lower animals.

K-Min Lee said there is a difference between nasal and temporal visual input. Drs. Lee, Norcia and Brodsky discussed the neuro-anatomy related to nasalward vs. temporalward stimulation of the retina.

Tina Rutar asked how does knowledge of neuro-anatomy help us understand what is taking place clinically?

Gerald Westheimer answered that a nasalward bias from the near reflex triad adds to the tonus that drives the eyes into adduction.

David Guyton said muscle length adaptation, which occurs with a large angle esotropia, leads to an even larger esotropia. Infantile esotropia is not necessarily due to increased esotonus.

Douglas Fredrick stated that perhaps infantile esotropia meets the criteria of cerebral palsy, characterized by increased tone and hyperreflexia.

Michael Brodsky said that latent nystagmus, POMA, and DVD/DHD arise from primitive visual vestibular reflexes, and these movements are normal in lateral-eyed animals. Does adding to one movement take away from another? Why do some patients only have latent nystagmus, only DVD/DHD, or POMA?

David Guyton mentioned that patients with infantile esotropia always have latent nystagmus. Is latent nystagmus the primary problem? Is infantile esotropia the result of convergence attempting to dampen/block the latent nystagmus? Patients with infantile esotropia may secondarily develop DVD/DHD. Is DVD also a mechanism to dampen latent nystagmus?

Alejandra de Alba Campomanes stated that if you fully neutralize the esotropia with prisms, you are more likely to detect DVD.

Michael Brodsky agreed, yes, the position of the eyes can affect what you see and what you call it. In a large esotropia, it is difficult to appreciate DVD and POMA.

David Romero Apis performed a maneuver of looking for DVD with total esotropia neutralized with prism. If you see DVD with this, you will have DVD post operatively.

Michael Brodsky commented on Dr. Simonsz' discussion of infantile esotropia. Cerebral palsy reflects a structural problem with the brain. You do not need a neurological insult to get infantile esotropia. Yet, sometimes if you straighten the eyes of an infantile esotropia patient, they will come back with a head tilt, and they also commonly have poor balance and are "clumsy." We have made the assumption that they are neurologically normal and merely have poor depth perception/binocularity. But do they suffer from another physiological/neurological imbalance?

Dr. Brodsky also commented on Dr. Guyton's suggestion that congenital nystagmus may be the primary problem in infantile esotropia patients. Congenital nystagmus can occur in isolation or as a secondary phenomenon; it is not always a primary visual problem. Normal monkeys given base out prism develop infantile esotropia.

ACCOMMODATIVE ESOTROPIA

3. Clinical Problem: How can we optimize the correction of hyperopia in the treatment of accommodative esotropia?

Why this is a problem: Not all strabismologists agree that accommodative esotropia should be treated with full plus refractive correction. Some children do not tolerate full plus refractive correction. Accurate retinoscopy in young children is difficult and depends on the skill of the examiner.

Scott Foster remarked that inadequate undercorrection of hyperopia can lead to recurrence of esotropia. If a child fuses, s/he will accept the full cycloplegic refraction. As the child ages, you no longer have to fully correct the hyperopia. Dr. Foster sees many consecutive exotropias in his practice, and he wonders whether some of these are accommodative esotropia patients who were inappropriately operated on.

Bradley Black explored the widespread practice of undercorrecting hyperopia in esotropic patients to promote emmetropization. Various studies have shown that the reduction in hyperopia in accommodative esotropia patients between 7-12 years of age is only small and does not differ between patients wearing the full cycloplegic refraction versus those undercorrected by one diopter. In his own retrospective study involving esotropic patients, who had a mean refractive error of +4.3 D initially, he was able to decrease plus over time but found no difference in the underlying cycloplegic refractive error in those corrected fully or partially.

He also addressed the question of whether accommodative esotropia resolves spontaneously. Thirty percent of patients with accommodative esotropia achieved resolution of esotropia by age 15 years. The best predictor of outgrowing accommodative esotropia was having had less hyperopia at presentation.

About 8% of accommodative esotropia patients present prior to age one year. This subgroup appears to have a low incidence of bifoveal fusion and a high incidence of deterioration eventually requiring surgery, thus overlapping with infantile esotropia. The older the child was at presentation with esotropia, the more likely s/he was to regain stereopsis.

Dr. Black responded to a comment made by Dr. Simonsz that the emmetropization process in children with accommodative esotropia may be different than the emmetropization among hyperopic children without esotropia.

Discussion:
A discussion ensued re the emmetropization process.

David Guyton commented that monkeys reared in the dark do not lose hyperopia.

Alejandra de Alba Campomanes said hyperopia increases in early childhood because of corneal flattening.

André Roth stated that refractive error at birth is +4.00 but diminishes rapidly to +1.75 at one year of age. If over +3.00, the best cycloplegic agent is giving the full refraction. (In other words, the best way to measure full plus is to perform refraction while the patient is wearing full plus.)

David Romero Apis said hyperopia increases at ages three to five years, and the only explanation is corneal curvature.

Huibert Simonsz mentioned that it is just statistical happenstance that some patients get more hyperopic and others do not over time.

Arthur Jampolsky remarked that E.V. Brown in the 1940s said hyperopia increases until age seven years. Dr. Jampolsky studied that corrected refraction can bring out latent hyperopia. It's not truly changing over time, but our ability to measure it improves over time. A good cycloplegic agent is habitually wearing the full cycloplegic refraction.

Tina Rutar mentioned that most of us are relatively comfortable treating the older child presenting with accommodative esotropia; we fully correct their hyperopia and the patient regains fusion/stereopsis. But how can we optimize the treatment of early onset accommodative esotropia (presenting at < six months of age)? Even when given full plus, these infants don't do well, rarely developing stereopsis and often developing features of infantile strabismus such as inferior oblique overaction and DVD.

Arvind Chandna questioned if you have an infant with the normal amount of plus for age but presenting with esotropia, would you give the full plus?

Scott Foster stated that in infantile esotropia patients, before he ever does a second operation, he re-refracts them and uses atropine.

Cameron Parsa said that in infants with Down syndrome, who hypoaccommodate, he gives full plus and full correction for near, so if the underlying cycloplegic refraction is +2.00, he prescribes +5.00 in single vision near spectacles.

Tina Rutar wanted to know if we should also prescribe full plus and the added +3.00 in single vision near spectacles to infants with early onset accommodative esotropia, whose visual world is at near?

John Sloper said to push the plus, but do not over plus.

Arthur Jampolsky further said to maximize the plus; with both eyes open (binocular), you can add +0.25 to 0.50 diopters over what is measured (monocular).

Jonathan Horton wondered if the child may not wear the spectacles if you push the plus.

Arvind Chandna said he gets relatively good results performing cycloplegic refraction after giving atropine 1% and waiting 90 minutes.

Stephen Kraft reminded us to use topical anesthetic before cycloplegia, which allows for better corneal penetration of the cycloplegic agent, and also makes it less likely that the child will cry out the second eye drop.

Tina Rutar said we all agree that we would like to measure the full underlying hyperopic refraction. What is the best way to achieve cycloplegia? Prescribing atropine for use for three days at home prior to the next clinic visit is difficult for the family and adds another visit. How can we improve?

    a.   Discover a new cycloplegic agent that is easier to use and safer than atropine?
    b.   Perfect imaging techniques that can calculate the refraction based on corneal curvature, effective lens position, axial length, and other biometric parameters?

Harley Bicas said that wearing the full cycloplegic refraction while using the atropine for three days will allow you to uncover the additional hyperopia. But wearing full plus glasses will interfere with emmetropization.

Bradley Black stated that it's an important clinical question: does wearing full plus actually impact the underlying hyperopia?

Douglas Fredrick said that myopes have much less accommodation, so you shouldn't compare myopes to hyperopes, who accommodate well. You have to take into consideration the anterior chamber depth and the lens position.

Henry Metz mentioned that the color of the skin and iris pigmentation are important. Darkly pigmented individuals can still accommodate even after receiving atropine.

Cameron Parsa questioned can we relax accommodation by using base out prism to relax convergence?

ESOTROPIA AT NEAR

4. **Clinical Problem: What is the best management strategy for esotropia at near (high accommodative convergence/accommodation ratio esotropia) in children who are already wearing their full hyperopic refraction, or who have negligible hyperopia?**

Why this is a problem: Young children may not wear bifocal spectacles or use them appropriately. Phospholine iodide has side effects and is difficult to obtain. Does surgery appropriately eliminate the excess esotropia at near?

Hermann Mühlendyck explained the Faden operation and compared it to the Demer posterior pulley fixation procedure as published by Clark et al. In his hands, he had the best effect with a Faden fixation at 14 mm rather than at 11 mm. In a Faden operation, one does not encounter a restriction on forced duction testing, but such a restriction is felt following the Demer procedure, caused by fixation of parts of the pulley sleeve.

Arvind Chandna described the clinical problem of managing esotropia at near. We should elucidate its cause: is it a problem of accommodation, convergence, or vergence, or combination of all three? Review of the literature does not give a clear entrance. The orthoptists define esotropia at near as fusion (stereopsis) at distance, with some fusional range, and esotropia at near. If the near esotropia goes away with a +3.00 add, prescribe a bifocal, otherwise, perform surgery. Dr. Chandna performs medial rectus loop recession for the near angle by hang back, allowing the middle of the muscle to go back further.

David Romero Apis said that a small bilateral medial rectus recession is not enough. A large conjunctival recession can help the near deviation.

28

John Sloper will try bifocals, otherwise bilateral medial rectus recession for near angle. Overcorrections at distance are rare.

Tina Rutar said that young children may not wear bifocals, or use them appropriately, especially if they are nearly emmetropic.

Michael Clarke noted that a large study in Germany found that these kids do get better with bifocals, but a natural history study was not done to see if they would get better on their own.

André Roth said that he gives a progressive bifocal, with the add set higher.

Alberto Ciancia determined convergence and divergence amplitudes with prism preoperatively. Response to surgery will depend on how strong fusional vergences are.

Arvind Chandna said that this is an important clinical entity to study: essentially, these patients have binocular function switched on for distance fixation and switched off for near fixation.

BINOCULAR VISION, AMBLYOPIA, AND SUPPRESSION

AMBLYOPIA ASSESSMENT

5. Clinical Problem: Can amblyopia be better assessed and monitored by using approaches other than measurement of high contrast visual acuity?

Why this is a problem: Most clinics are not equipped to measure contrast sensitivity, grating acuity, Vernier acuity, VEPs, etc. Subtypes of amblyopia differ in their effects on these parameters. Visual acuity testing is unreliable in young children.

William Good remarked that VEP measurements are a basic research tool that is helpful in understanding cortical processing. VEP can be used to monitor patients longitudinally, and provides reliable interocular differences.

Dr. Good described the contrast sweep, grating sweep, and Vernier sweep VEP. All three functions take 10-15 minutes to run per patient. He is currently using these tests to study the effects of neonatal hyperbilirubinemia on the visual cortex. The limitations of VEP include the time it takes to master the test, the cost of the equipment, and the need for a good assistant to run these tests in a busy practice. Currently, there is also no simple algorithm for interpreting test results.

The sweep Vernier VEP, in which a sweep of different sized Vernier line offsets is presented to the patient, correlates well with optotype acuity. The sweep Vernier VEP can be done at a very young age. This VEP test and others might be ready for clinical prime time.

Vernon Odom said that VEP is a great research tool, but it is not very useful in regular practice. One needs a good research assistant, and the cost and time involved are not reimbursable.

Arvind Chandna remarked that optotype acuity is not the only thing we should be looking at when measuring amblyopia. Grating acuity, Vernier acuity, contour integration studies, and contrast sensitivity tests can give us more information. These measures will differ in different subtypes of amblyopia.

K-Min Lee said that when assessing vision with the amblyopic eye, what stimulus should be simultaneously presented to the sound eye? Patching the sound eye is not a real world situation. Testing visual function in the amblyopic eye under truly binocular testing conditions will give different results.

Anthony Norcia stated that it is what you would expect. Stimulating the dominant eye shuts down the nondominant eye because the nondominant eye is being suppressed. If the nondominant eye is stimulated, there is no impact on vision in the dominant eye.

In treating amblyopia with patching, you are treating multiple visual "systems" and you could over patch one system and under patch another. Different areas of the visual cortex have different developmental plasticities.

Henry Metz questioned that isn't optotype acuity an important clinical indicator for determining whether to continue amblyopia treatment?

Arvind Chandna said that visual functions other than optotype acuity may improve with amblyopia treatment, and this improvement won't be detected if optotype acuity alone is clinically used.

A patient with a monocular cataract may be functional with the dominant eye patched even though optotype acuity is terrible.

Anthony Norcia noted that Vernier acuity can be measured in a three-month-old child, and it continues to develop through

31

puberty.

Arvind Chandna said that anisometropic amblyopes who are treated with spectacles alone improve in optotype but not in Vernier acuity.

Harley Bicas said that the most important tests in the assessment of amblyopia are simple clinical tests: the cover test, reaction to patching, refraction, and ophthalmoscopy.

William Good used sweep VEP techniques to detect amblyopia in some patients with unilateral infantile eyelid hemangiomas, who clinically had no evidence of amblyopia. They had no strabismus and no induced anisometropia. He wonders whether occlusion of the visual axis in eccentric gaze positions might be driving amblyopia, even when there is no occlusion in primary position.

Stephen Kraft clarified that current VEP techniques involve extrapolation to determine the achieved threshold. Performing VEP testing in the frequency domain (presenting data at a particular frequency and seeing if there is a cortical response at the same frequency) would eliminate that extrapolation and give you more accurate information.

Anthony Norcia/Vernon Odom commented that that would not necessarily be more accurate.

Alejandra de Alba Campomanes said that why not just use Teller cards to quantify visual function in preverbal children? Teller cards were used in the esotropia retinopathy of prematurity (ROP) study. Why? Because there were multiple centers involved, they are less expensive that VEP testing, so that visual results could be more directly compared to older ROP studies?

William Good answered that Teller grating cards and sweep grating VEP overestimate optotype acuity. Sweep Vernier VEP has a better correlation with Snellen optotype acuity.

Federico Vélez said that Teller acuity is so variable. Is VEP more reproducible?

William Good answered that VEP is more reproducible. Unlike other clinical measures of visual function, which are subjective, VEP is objective.

AMBLYOPIA TREATMENT AND CORTICAL PLASTICITY

6. **Clinical Problem: Amblyopia treatment is primarily limited to refractive correction, occlusion, penalization, and correction of strabismus. Treatment is most effective when instituted at a young age. How can we improve upon these limited treatment options?**

Why this is a problem: Amblyopia may improve but not resolve, especially with late diagnosis. Treatment compliance is poor.

Cameron Parsa discussed problems with the recent studies suggesting levodopa and acupuncture may improve cortical plasticity/responsiveness to amblyopia treatment. Levodopa is a neurotransmitter that is also active at the retinal level. (In Parkinsonism, contrast sensitivity decreases). In the levodopa studies, visual acuity gains were seen in both eyes. Perhaps levodopa is enhancing retinal sensitivity, not cortical plasticity. Similarly, acupuncture may induce an adrenergic response that heightens retinal sensitivity.

He also cautioned about the possible adverse side effects of levodopa. What is happening to local dopaminergic regulation in patients receiving exogenous levodopa? We also need to recognize that different subtypes of amblyopia: deprivation, strabismic, anisometropic, may have different periods of cortical plasticity.

John Sloper urged extreme caution in treating teenagers with levodopa, which is a neurotransmitter important in many central nervous system functions. What are we doing to other areas of the developing teenagers' brain? What will be this patient's risk of developing schizophrenia? The immediate effects of levodopa include dyskinesia and serotonin depletion (which can result in depression). These effects limit the use of levodopa in adults with Parkinsonism.

Stephen Kraft agreed that long term use could have systemic adverse effects, but doubts this will happen from short term use lasting weeks to months, as described in current studies on the use of levodopa for enhancing cortical plasticity in amblyopia treatment. However, he acknowledges the lack of long term follow up to know the long-term safety profile of this medication.

Arvind Chandna mentioned that after discontinuation of levodopa, there is regression in vision gains to the level before drug treatment was begun.

John Sloper said that even one dose of a psychotropic drug can have permanent effects.

Anja Palmowski-Wolfe reviewed the literature on adult patients demonstrating gains in vision in amblyopic eyes. In 1957, Carl Kupfer showed that 8 adult patients who lost the sound eye for a myriad of reasons had recovery of vision in the amblyopic eye. She reviewed the literature from Uri Polat on perceptual learning, suggesting that the visual cortex remains plastic into adulthood. There is fMRI data in support of perceptual learning.

Denise Satterfield wondered that could these patients simply be learning how to perform the test better over time?

Anja Palmowski-Wolfe said that after perceptual learning exercises using Gabor patches, the patients' amblyopic eyes improved in visual acuity and contrast sensitivity, not on measures of Gabor patch recognition.

Suzanne McKee added that if it were just a matter of learning the task, then the good, untrained eye would also get better. Video game-based training programs work to improve visual acuity in adult amblyopic patients better than patching does.

Arvind Chandna wondered if these improved adult amblyopic patients represent regressed amblyopes? They will get better with reinstitution of patching.

K-Min Lee suggested that the timing of stimulation to the amblyopic eye might be important. Perhaps binocular neurons can boost the stimulus getting to the amblyopic eye a little earlier? He brought up an analogy related to delays in input in patients with multiple sclerosis and optic neuritis, where the delay can create problems with perception. When the delayed eye is stimulated a little bit ahead of time, big changes occur in the VEP.

Stephen Kraft said that the recent Agnes Wong article in the *Canadian Journal of Ophthalmology* covers a lot of information on cortical plasticity and perceptual learning, including fMRI data.

Arthur Jampolsky suggested that Agnes Wong was apparently unaware of the older orthoptics literature.

John Sloper cautioned on the interpretation of fMRI as there is a lot we do not understand yet. For instance, magnocellular stimulation appears to give more blood oxygen depth than a parvocellular stimulus. Thus, fMRI is very dependent on the stimulus used.

K-Min Lee said that the orthoptists, in the older literature, did not have the ability to modulate the timing of binocular inputs at the level we can today. We now have good computerized ability to modulate the stimuli with respect to delivery time and stimulus type. Thus, we can deliver binocular stimuli within the tolerance of fusion, for example 20-30 msec apart, maintaining perceptual simultaneity with asynchronous delivery of stimuli.

Felisa Shokida asked whether strabismus surgery can improve binocular cortical activation in adult patients. She showed pre- and post-op fMRI data; eight of 10 adult patients who underwent strabismus surgery showed increased binocular cortical activation on fMRI. Two of 10 patients, one with Ciancia syndrome and one with dissociated horizontal deviation (DHD), showed decreased activation.

Omondi Nyong'o commented that adult strabismus patients improve in quality of life scores after strabismus surgery. Mayo uses the validated AS20 questionnaire.

John Sloper said that the way the questionnaire is designed is important; some questionnaires focus on asthenopia complaints. Some quality of life questionnaires will always show an improvement after intervention.

Lora Likova stated that treatment of amblyopia in older patients is an unsolved clinical problem that will require understanding of adult brain plasticity. She performed a study involving congenitally blind subjects. They performed a cognitive kinesthetic training task. Under total visual deprivation, they tactilely explored images and memorized them for 20 seconds. They were then asked to draw the images. Sample images included those of a profile of a face and a boot. At first, this was a task impossible to master for either sighted or blind subjects. However, after a week of "training," blind subjects were able to draw images recognizable to the actual kinesthetically explored objects. Lora Likova demonstrated a change in fMRI activation in the primary visual cortex in one such congenitally blind person.

Arvind Chandna was stunned by the data as these were congenitally blind subjects. So how can this be explained apart from some form of visual memory?

K-Min Lee asked Dr. Likova if her subjects could imagine visually?

Lora Likova said that she is unable to ask because they are congenitally blind. The literature supports visual imagery causing weak activation of V1. One hypothesis that may explain the data is to suggest that the brain, V1, has an amodal spatial sketchpad rather than a visual spatial sketchpad. Thus space is beyond any separate modality and can be perceived texturally, visually, mathematically, or even auditorally.

Tina Rutar wondered why must the stimulus be emotionally interesting?

Lora Likova said that training is best accomplished using real-life tasks, which are attentionally and emotionally stimulating. It is not simply perception/action but rather a perception/cognition/action loop, which is how we solve real-life problems.

John Sloper questioned what is the input channel? Is it still the lateral geniculate nucleus (LGN)?

Lora Likova said that that is an important unanswered question. This could result from bottom up or top down processing.

John Sloper said that Tony Murray studied infantile esotropia patients in South Africa who had never been treated. 75% did NOT have amblyopia.

Christopher Tyler asked what is motor fusion?

John Sloper answered that motor fusion and sensory fusion usually go together, but they can be separated. He gave the example of a patient who could fuse to 50 seconds of arc fleetingly but lacked motor fusion to hold it there and consequently had double vision most of the time.

SUPPRESSION

7. **Clinical problem: Suppression is friend and foe. Amblyopia treatment often involves breaking down suppression that has developed as a compensatory mechanism to avoid diplopia and visual confusion.**

Why this is a problem: We could use a better understanding of suppression to our advantage in treating amblyopia.

Richard Harrad asked -- can you see through your nose? Yes, you can. You suppress the image of your nose and think you see through your nose with the adducting eye. Suppression is happening all of the time and is a normal adaptive mechanism. We use suppression to avoid physiologic diplopia.

Suppression takes on different forms in anisometropic and strabismic amblyopia. The amblyopic anisometropic eye is suppressed independent of whether it is a crossed or uncrossed disparity. In esotropic amblyopia, however, there is suppression only of the uncrossed disparities. For the crossed disparities, the amblyopic eye can actually reduce the response in the sound eye.

How strong is suppression? What is happening to suppression when we treat amblyopia?

Suzanne McKee/Richard Harrad showed dichoptic masking in anisometropic amblyopes. In strabismic amblyopes, suppression is more powerful than one would expect on the basis of the difference between the contrast sensitivity of the two eyes.

K-Min Lee wondered what are differences between suppression and binocular rivalry?

Suzanne McKee answered that amblyopes can experience

binocular rivalry. A paper by Cliff Shore discusses this.

Arthur Jampolsky commented that binocular rivalry is when a single percept alternates between the two eyes for corresponding competing loci. Bielschowsky said that binocular rivalry was the first step to suppression.

John Sloper said that patients with Duane syndrome are interesting. Those with better head control get less suppression than those who have worse head posture control.

Henry Metz stated that some post head trauma patients can no longer suppress physiologic diplopia.

Christopher Tyler explored what it is like to not have stereopsis by inducing a vertical diplopia.

Stephen Kraft stressed the importance of good definitions and terminology in measuring suppression clinically. Some tests are more dissociating than others (for example, Worth 4 dot compared to Baggolini tests). Patients are inconsistent in their responses from one measurement to the next. There are no good clinical predictors re: depth of suppression, or the extent of suppression in the visual field.

We need better ways to evaluate our patients. Factors influencing suppression are:

a. Dissociating ability of test.
b. Is the test detecting central or peripheral suppression mechanisms?
c. Are patients really suppressing, or are they subjectively ignoring the image?
d. Age at onset and length of time of strabismus. It may be logical to expect the longer the strabismus, the deeper the suppression gets. However, patients who develop an intermittent exotropia can have profound suppression in their dissociated situation within a

very short amount of time.
e. The angle of strabismus.
f. Asymmetry of sensory inputs (size, acuity), e.g. will a difference in image size between the eyes drive deeper suppression?
g. Are these children normal cortically?

We have more questions than answers. We must better define how to measure depth of suppression.

Henry Metz said that it is also important to look at the extent of the suppression; does it affect the fovea or only the periphery?

Arthur Jampolsky asked why do you measure suppression? A lot of data is taken regarding suppression and systematically ignored in making clinical decisions. But you can predict some things based on a simple algorithm. If you make an exo into an eso, you will get diplopia. If you have an adult eso and can put prism up to ½ the measured angle without diplopia, you won't get diplopia. So why do all these tests?

John Sloper said that suppression is his friend. Do you need to get rid of suppression to get good binocular vision? No, the mechanisms of suppression are not the same as mechanism of fusion. Suppression and binocular vision can coexist.

He uses prisms to attempt to predict if the patient is at risk of postoperative diplopia. If there is some risk, He uses Botox rather than surgery as the initial intervention.

Richard Harrad described research in which one balances the contrast of inputs into each eye to the point that the amblyopic eye can see the stimulus. For example, a child can watch a videogame that is presented to both eyes simultaneously, but with differing contrasts balanced to compensate for the amblyopia. Then, one can develop facilitatory binocular interactions. They are not breaking down suppression. Hopefully, such patients will not develop intractable diplopia.

John Sloper said that there are people with a normal sensory binocular framework who have longstanding strabismus, and you can recover normal binocular vision in these individuals. But they are few.

Keith McNeer said that suppression won't develop if infants achieve orthotropia early through the use of Botox.

Hermann Mühlendyck stated that suppression exists primarily to avoid visual confusion, which is far more disturbing that diplopia.

DIPLOPIA

8. **Clinical Problem: When treating adults with strabismus, we cannot accurately predict who is at risk for postoperative diplopia, nor its severity.**

Why this is a problem: The fear of postoperative diplopia leaves some adults with uncorrected strabismus, and thus they do not experience optimal binocular function or the psychosocial benefit of straight eyes.

David Romero Apis said that he measures the size of the suppression scotoma before surgery with a prism bar. Ask the patient when they begin to see double.

Alejandra de Alba Campomanes mentioned that if Dr. Jampolsky states that intermittent exotropia patients with equal visual acuity will regain motor fusion if you get the eyes straight, what then determines optimal timing of intermittent exotropia surgery?

Arthur Jampolsky answered that he operates on intermittent exotropia if the frequency increases, the lateral rectus tightens, or the obliques become involved.

Michael Clarke said that in England, it is law that orthoptists have to do a post-operative diplopia test!

Scott Foster said that there are patients with large exotropias who don't like being straight and will complain of diplopia.

Bradley Black said that he has hardly ever had to put someone back into an exotropic position surgically.

John Sloper stated that diplopia is attention-dependent.

9. Clinical Problem: How can we better manage intractable diplopia in adult patients? Can adults be taught to suppress?

Why this is a problem: If we better understand suppression, we could use that understanding to our advantage in treating amblyopia. We could also teach adults to suppress when necessary to relieve diplopia or visual confusion,

Denise Satterfield said that there are instances when we cannot restore fusion. Patients will send her images illustrating their diplopic visual world. There are many reasons for intractable diplopia, including torsion, oscillopsia, aneisokonia, and others. In addition to monocular occlusion and monocular blur, we can try monovision for these patients.

Jonathan Horton said that we work hard to preserve binocular vision and some cataract and refractive surgeons intentionally cause monovision, precipitating loss of binocularity and diplopia.

David Guyton remarked that he finds it takes much more induced anisometropia than used in monovision to ignore the second image, up to 15 diopters.

Arthur Jampolsky noted that an old option by Arthur Linksz was to put a small spot of colorless nail polish on the central portion of the spectacle lens. The periphery is much more forgiving.

Stephen Kraft said that the amount of anisometropia depends on which eye you use for fixing at distance and near. It's better to set the dominant eye for distance and the nondominant one for near.

David Romero Apis said he finds it is helpful to predict who will have postoperative diplopia. Neurotic doctors perform these tests on neurotic patients.

Henry Metz remarked, yes, some people complain more than others, some are tolerant of diplopia.

Federico Vélez said that for monovision to work and not disrupt binocularity, it cannot be more than 2D. More work is needed to understand at what point monovision breaks down the binocular system and causes strabismus and diplopia. Sometimes, it's difficult to tell which eye to set for distance and which for near.

Alberto Ciancia said that you can also apply satin scotch tape in an area of the spectacles to eliminate diplopia.

Stephen Kraft said that you can also use Bangerter filters. Use the clearest one that still eliminates the diplopia.

David Guyton described the light on/off test for the detection of dragged fovea diplopia syndrome, a cause of intractable diplopia. There is binocular central diplopia in presence of overall peripheral fusion. This usually occurs in patients with visual acuity of 20/50 or better, who have a central epiretinal membrane.

Prisms don't work; refractive blur rarely works; total occlusion is not practical; but partial occlusion with satin scotch tape will work.

Cameron Parsa said that nearly always, these patients have vertical diplopia because Panum's fusional space is so restricted in the vertical meridian.

John Sloper asked Dr. Guyton if he has tried the Jampolsky suggestion of giving a ring of occlusion in order to leave only a small central area for fusion?

David Guyton said that it doesn't work. Some people with dragged fovea diplopia syndrome have difficulty driving. He likes using a single vertical strip of tape over the distance portion of one lens because it leaves the side vision open when driving so they can see things coming at them from the side. Others just have trouble reading, in which he uses a dab of fingernail polish with a little tape over the bifocals and so when they look down, it blocks that eye. You don't have to put tape over the whole lens or even over the whole vertical strip.

INTERMITTENT EXOTROPIA

10. Clinical Problem: Surgical results in intermittent exotropia are unstable. How can we better determine the timing and the type of intervention?

Why this is a problem? Exotropia frequently recurs, and patients require reoperations. Overcorrections in patients who are young can cause suppression and development of amblyopia, and overcorrections in adults result in diplopia.

William Good provided examples of clinical patients illustrating intermittent exotropia as a vexing and recurrent problem. For example, acting on the advice of Creig Hoyt, to "overcorrect these patients by a lot" in the intermediate postoperative period, Dr. Good operated on a child with a 30 PD exotropia, had a consecutive 30 PD esotropia, but two years later, the child was again orthotropic. Another three years later, the intermittent exotropia had recurred.

Anja Palmowski-Wolfe summarized Dr. Hoyt's editorial on early versus late surgery for intermittent exotropia. Factors to consider in the timing of surgery include the presence of stereopsis at near in the vast majority of these patients, the low rate of amblyopia, and the uncertainty about the natural history.

André Roth discussed that his ideal timing for surgical intervention is a compromise, and that surgery at age four to five years may be the appropriate compromise. Before four years of age, the tonic vergence defect may not be completely manifest. To wait longer than four to five years of age would compromise potential normal binocularity. He also prefers that these criteria be met preoperatively - (a) both eyes must alternate, (b) provide full (Fresnel) prism compensation preoperatively to increase the power of fusional convergence, and to determine the maximum angle of deviation, and (c) measure the basic exotropia angle in the far distance (50 m).

Stephen Kraft questioned why not give less than full prismatic correction in a Fresnel prism preoperatively in order to leave some fusional convergence?

André Roth said that relaxing the motor compensation by accommodative convergence completely, and then the fusional convergence can be improved.

Federico Vélez said he performs a prism adaptation test on intermittent exotropia in the clinic, but leaves the Fresnel prism on only for an hour. He finds that the angle increases by 10-30 PD, sometimes even doubling.

André Roth said the reason for the preoperative Fresnel prism adaptation period is not only to find the maximum angle of deviation, but also to train the fusional vergence system so that the eyes will become straight rapidly after surgery.

Scott Foster discussed two subtypes of patients we may see clinically presenting with exotropia, which he finds rather straightforward in terms of management. One is the recurrent exotropia patient who was an accommodative esotrope treated surgically rather than with full correction. These patients have small adduction deficits and do well with bilateral medial rectus advancement. A second type of patient is the adult with intermittent exotropia. Dr. Foster will put prisms in front of the patient to determine what amount of prism brings out diplopia, and then operate on that angle, which is less than the total angle. For example, the patient intermittently fuses a 50 PD exotropia but reports diplopia with greater than 35 PD BO; operate on 35 PD leaving a 15 PD exotropia that the patient will easily fuse due to large convergence amplitudes.

The unpredictable patients are the children with intermittent exotropia. Some recur as early as six weeks postoperatively, whereas others remain straight forever. We don't know how to preoperatively determine who will do well and who will not. Patching to bring out the maximum deviation in these patients seems to affect only the near, not the distance angle. And the type of surgery, whether it is a recess/resect, or a bilateral lateral rectus recession, seems not to matter. Both in his personal experience and review of the literature, they all seem to have equally bad results.

Dr. Foster suggested a research study by Smith-Kettlewell and its colleagues on the use of the prism adaptation test in intermittent exotropia. Several studies from abroad have already shown that operating on larger angles, brought out by a one to two hour Fresnel prism adaptation test in the office, leads to better outcomes than operating on smaller angles from non-prism adapted patients. A larger study from multiple centers, however, would be beneficial.

Federico Vélez wondered, in response to discussion regarding the duration of the prism adaptation test – can it be done in one or two hours in the clinic, or should it be done for weeks? He mentioned his experience and that published in the literature on prism adaptation tests for esotropia. One or two hours of prism adaptation in clinic is sufficient for esotropia patients.

Michael Clarke discussed natural history data on intermittent exotropia in children from several centers in the United Kingdom. Most tend to volley back and forth between different levels of intermittent exotropia control; few resolve and few deteriorate into a truly constant exotropia.

Although we eagerly await data from the Pediatric Eye Disease Investigator Group (PEDIG) study on patching in intermittent exotropia, his impression on current data and experience is that overminus glasses and patching are not very effective treatments. Surgical treatment of intermittent exotropia provides short-term success in 2/3 of patients, but poor outcomes (persistent overcorrections and recurrences of exotropia) in 1/3 of patients. The longer you follow the patients, the more recurrences you tend to see.

Dr. Clarke discussed his experience designing and executing a randomized controlled trial on surgical treatment vs. active surveillance of intermittent exotropia. Two hundred thirty one patients were screened but recruitment was poor, with only 49 patients randomized. Some of the difficulties in recruitment included (a) inability to obtain baseline stereoacuity data in young children, (b) control scores too good to meet criteria for randomization, and (c) the inability to convince parents to allow surgery on what a parent perceived as a minor problem, or to allow monitoring instead of surgery for what another parent perceived as a serious problem. This study is a pilot study, and information learned from it will be used to plan a larger scale randomized controlled trial in the future.

Studying early vs. later surgery is also important. John Pratt-Johnson first suggested that early surgery may offer better long-term results. However, compounding the difficulties mentioned above in designing a randomized controlled trial will be the difficulty in obtaining baseline stereoacuity data from children under age four years. Stereopsis is still developing at that age, and you need to have a reliable baseline measure to serve as the basis for comparison.

In the short term, surgery offers good results. We do not know the ideal timing of surgery, nor do we know whether surgery at a young age would protect against reoperations down the road. Measuring clinical outcomes in children operated on prior to age four years is problematic because stereopsis is still developing at that age, and some patients may not have reliable baseline/preoperative assessments.

Huibert Simonsz wondered what are the primary outcome measures going to be?

Michael Clarke said that the outcome measures will be (a) exotropia control score, (b) alignment, and (c) stereoacuity, as well as a composite outcome measure.

Keith McNeer said that he uses Botox injections 2.5 units into the lateral rectus muscles for the treatment of intermittent exotropia, preferably at age three years. On follow up six years later, 21% of children had no deviation and 70% had small angle exophorias of approximately 10 PD that they could easily refuse. There were no consecutive esotropias. Dr. McNeer argues that Botox is a safe and superior alternative to surgery, and that the goal for these patients should not be elimination of any exophoria, but comfortable control of small exophorias.

John Sloper stated that all these children are different in terms of angle as well as control, and the decision on whether and when to operate should be made on an individual basis. He frecently operated on two kids six months of age who were manifest 90% of the time and were clearly going to get worse without intervention.

Dr. Sloper also presented data on intermittent exotropes (but excluding those with convergence insufficiency), showing that there is really no clear-cut way to categorize patients according to distance/near disparity. The distance angle, by definition, is greater than the near angle. However, for a given distance angle, there is a wide distribution of near angles, and no clear clustering into groups (such as divergence excess D>N or basic type D=N).

Patients also vary widely in terms of level of control. Level of control doesn't correlate with measures of stereopsis (at near) or of angle. Surgeons differ in terms of their criteria/comfort on operating for intermittent exotropia. Those who will generally operate only on patients who have worsening control over time, rather than those with stable control, may be reporting worse surgical outcomes because they are self-selecting patients who have a higher likelihood of deterioration.

Should we overcorrect intermittent exotropes and

by how much? Dr. Sloper presented data suggesting that the amount of short-term overcorrection does not influence postoperative alignment at five years. Furthermore, patients who were overcorrected had no more surgery per millimeter than those who were orthotropic or undercorrected.

Jonathan Horton remarked that it defies logic, that surgical dose does not correlate with effect.

Huibert Simonsz said that even in a well-controlled study, in which surgical dose per angle of deviation is strictly defined, there are variances in treatment effect. Furthermore, he was involved in a trial in which, no matter how strictly the protocol tried to "force" surgeons to follow a particular surgical table, 40% of them deviated from the protocol. Surgical dose is like religion to these surgeons, he said.

John Sloper stated that the three laws of thermodynamics apply well to intermittent exotropia: the first law is that you can never win; you can only break even. The second law is you can only break even at absolute zero. And the third law is that you can never reach absolute zero.

Arthur Jampolsky remarked that orthoptists in England taught us about the use of overminus glasses in intermittent exotropia. A small amount of overminus, say 0.5 to 1 D, is very helpful in the postoperative period if there is a suggestion of recurrent exotropic drift. He thought it was Harold W. Brown who first offered the notion that if you keep the eyes aligned, then the muscles adapt in length to that alignment. Defeat bad habits and encourage good habits.

John Sloper said that certainly overminus glasses improve the control in the short to medium term. The question is whether you can actually wean the patients out of the over minus lenses and maintain control. There is also research from orthoptist Anna Horwood suggesting that fusional vergence rather than accommodative vergence is responsible for control of intermittent exotropia.

David Romero Apis described that there are two types of intermittent exotropia patients who look identical from the motor perspective, but have a very different sensorial background. One is the true intermittent exotrope, who has normal stereopsis. The other is the pseudo intermittent exotrope, who has poor stereopsis, and unusual diplopia due to unharmonious and anomalous retinal correspondence near the neutralization angle. This first type of patient will not remain overcorrected in the long term, but the second type of patient, if overcorrected, will remain an esotrope forever.

David Guyton said he performs adjustable suture surgery for children with intermittent exotropia, using a second dose of propofol anesthesia in the postoperative area. An analysis of his own data showed an 11% improvement with use of adjustable sutures, but he still could not determine the appropriate target treatment angle.

Cameron Parsa provided an example of a medical student who has intermittent exotropia with relatively good control and 40 arc seconds. The student was overcorrected by 25 PD with a bilateral lateral rectus recession, and the consecutive esotropia slowly diminished over time, such that one year out, the student had intermittent diplopia only at distance fixation. Now, three years later, the student is exotropic again. Here is an adult with fusional divergence and the ability to suppress the hemiretina. How could the patient been more optimally treated?

Should we operate on intermittent exotropes early in life and deliberately make them into monofixational esotropes? So the question is, do we want high-grade stereopsis so badly that we will tolerate multiple reoperations? Or should we achieve stable results at the expense of stereopsis.

Jonathan Horton thought that trading esotropia for exotropia is a bad idea.

Federico Vélez said that Stacy Pineles re-examined patients 10 years after surgical treatment for intermittent exotropia. If your outcome is simply motor alignment, 70% of patients were doing well. But if you defined a good outcome as good sensory fusion plus good motor alignment, the long-term success rate data drops to 30%.

John Sloper said that some of these patients had more than one surgery.

Federico Vélez answered, absolutely, but the results are the same. In another paper that we published in the *Journal of AAPOS*, we looked at all patients who were esotropic at near and distance on postoperative day one and followed them for at least four years. The majority became ortho or exotropic. There were no significant differences when you compared patients by type of surgery (unilateral recession/resection or bilateral lateral rectus recession) or type of intermittent exotropia.

David Romero Apis remarked that the final angle depends on the sensorial status of the patient. Patients with a good sensory status will not remain overcorrected.

11. Clinical Problem: Inadequate understanding of the underlying cause of intermittent exotropia may be handicapping our treatment methods and success.

Why this is a problem: Understanding the underlying cause of intermittent exotropia may lead to better treatments.

Jonathan Horton described his research study on suppression in exotropia, in press in the *Journal of Neuroscience*:

We have mapped suppression scotomas in patients with strabismus using dichoptic visual field mapping procedures, and then we've raised monkeys with exotropia and looked at the pattern of metabolic activity in their visual cortices. We found a pattern of metabolic activity that maps the pattern of visual field suppression, obtained psychophysically pre-morbidly.

The first study involved a cohort of human subjects who had longstanding exotropia, ability to alternate freely between the two eyes, and 20/20 visual acuity in each eye. We used red filter over the right eye, blue filter over the left eye, had them fixate on a central point, and then we flashed purple stimuli. The catch trials were red and blue, answers to those trials are unambiguous, but then there were purple stimuli consisting of isoluminant red or blue and if the subject said red, we knew they saw that part of the visual field with the right eye, and if they said blue, we knew they saw it with the left eye.

Dr. Horton found (surprisingly) that the fovea of the deviated eye is perceptually active. It is not suppressed. There is a demarcation line between the projections of the two foveas onto the tangent screen, where perception switches from one eye to the other. The amount of temporal retina that is suppressed depends upon the angle of the deviation. In subjects that have a very, very large angle of deviation, the amount of temporal retinal suppression is correspondingly less. The smaller the arc of deviation, the closer to the midline of the retina the suppression extends.

Next, Dr. Horton schematically diagrammed the projection of this pattern of visual field perception onto the visual cortex in each hemisphere of the human.

He said that the advantage of working with the macaque monkey is that we can look directly at the anatomy of this and see if the pattern of metabolic activity we see corresponds to what we predict. Four macaque monkeys were raised exotropic by cutting their medial rectus muscles when they were about a month old. At age three to four years, these animals alternate freely between the two eyes. They have good acuity in both eyes. They have a pattern of visual field suppression that matches what we find in humans when we map them dichoptically.

We then sacrifice the animals and dissect the striate cortex, lay it down so it is completely flat and then we section it. And we process the tissue for cytochrome oxidase, a metabolic enzyme that gives us a sort of average of cortical activity over the days that preceded the animal's death. In order to determine which eye "owns" a particular ocular dominance column in the striate cortex, we also injected a radioactive tracer into the right eye, which is transported transsynaptically up to the cortex, just before sacrificing the animal.

The main result was that there was equal metabolic activity in the middle of the ocular dominance columns serving the left and the right eye. There was also reduced cytochrome oxidase activity and thus reduced metabolism in the ocular dominance columns serving the peripheral temporal retina of the ipsilateral eye, which is a finding that jibes with the psychophysical studies of suppression scotoma in exotropia.

These studies provide a metabolic fingerprint of visual

suppression that we are picking up in the primary visual cortex of monkey. He thinks this represents an advance in our understanding of what is going on at a fundamental level with visual suppression in the condition of exotropia.

Arthur Jampolsky clarified that Dr. Horton's research paradigm here, had foveas with dissimilar stimuli - though simultaneously. These data were acquired using dichoptic stimulation, meaning you have a different image in the right eye and a different image in the left eye. The very essence of binocular vision is simultaneity of similar stimuli (similar contours). If dissimilar stimuli are used, it simply is not a binocular vision paradigm circumstance.

Jonathan Horton defended his study design and described how an individual can have both foveas active at the same time in exotropia:

Obviously that conjures up the problem of visual confusion, and what we found when we did after imaging testing in these patients is that they have anomalous retinal correspondence. Anomalous retinal correspondence is absolutely key to understanding the sensory adaptation that occurs in exotropia. They remap this deviated eyes fovea in their body axis in their parietal perception of where objects are laid out in their personal environment, and they have a subsidiary status assigned to that deviated eye. It doesn't enjoy the same type of visual attention and visual perceptibilities as the fovea of the eye that is being used to fixate the required targets, but nonetheless they perceive land in the vicinity of that deviated eye's fovea.

We have looked at the ocular motor behavior of these alternating exotropes and, as they do the visual scene, the decision about which eye to use to acquire the next target is made by seeing the next target in the retina of the eye that action makes the saccade to the next target. Often it's in the peripheral retina of the deviating eye within immediate movement of that deviated eye's fovea on to the target of interest. That fovea has not been suppressed in its deviated position.

David Guyton clarified the study design, that perhaps the subjects see a small central target only?

Jonathan Horton described the appearance of the screen that the subjects view, which has a stimulus that should also be perceived by the peripheral retina.

Michael Brodsky investigated the role of luminance in intermittent exotropia. These children are unusually sensitive to light and they close one eye in bright sunlight. To my mind, there have always been three possibilities. (a) They are abnormally sensitive to light. A lot of photophobia has recently been related to melanopsin and the retinohypothalamic tract pathways. (b) Could increased luminance cause a divergence movement prompting closure of the nondominant eye so that the dominant eye doesn't drift exo? (c) What if there is light-induced esotonus and they get momentary diplopia and that is why they close one eye?

Using three-dimensional video-oculography, we recorded eye movements under a variety of circumstances. We could not bring out any evidence of light-induced exotonus or esotonus. In darkness over two minutes, the eyes do not move. Some go in a little bit, some go out a little bit, but on average they stay completely straight in intermittent exotropes. So there is some tonus that is actively holding the eyes straight even in intermittent exotropes. In darkness if you have them look at tiny fixation device and you cover one eye and then you cover the other eye, the eyes will go out. It seems that you need fixation in intermittent exotropia to drive the deviation. If you think about it, you never see an intermittent exotrope, where they zone out for a minute and both eyes go out; it only happens when they are fixing with one eye. We have seen these DHD patients where - Michael Graef calls it the 13 X 17 test - you say what is 13 x 17 and they start thinking about it and both eyes go out. But, in an intermittent exotrope you don't see that.

We looked at the melanopsin pathway in a few patients with intermittent exotropia. It's an attractive pathway to look at because it is subcortical, it's luminance, and the melanopsin pathways mediate photalgia and patients with intermittent exotropia have that. However, we cannot find any melanopsin effect associated with intermittent exotropia.

Jonathan Horton wondered whether accommodation explains the lack of outward drift among intermittent exotropes in total darkness. When subjects are put into the dark, they naturally accommodated at a point about a meter in front of them.

Michael Brodsky stated that he should look at the size of the pupils of these subjects to look for pupillary constriction.

David Romero Apis described the clinical differences between intermittent exotropia and DHD.
  a. Intermittent exotropia is symmetrical; DHD is not.
  b. Intermittent exotropes have times of fusion without suppression, whereas patients with DHD suppress all the time, even when the eyes appear straight.
  c. Intermittent exotropia is either present or absent and when present, a constant angle, whereas the angle of DHD changes from moment to moment.
  d. DHD disappears when both eyes are covered, but the outward drift of one eye in intermittent exotropia persists when both eyes are covered.
  e. Occlusion nystagmus is seen with DHD, not intermittent exotropia.

Dr. Romero Apis has never seen a case of intermittent exotropia coexisting with DHD.

Discussion:
There was discussion on whether intermittent exotropia and DVD could coexist.

Stephen Kraft said that monofixational intermittent exotropia is less common that monofixational esotropia, but such cases might explain some of the confusion in the published studies purporting intermittent exotropes (miscategorized monofixational intermittent exotropes) can manifest DHD.

Alberto Ciancia described additional features that distinguish intermittent exotropia and DHD, including the presence of latent nystagmus and vertical movements in the latter condition. He also has seen infantile esotropia patients with dissociated DHD: you cover one eye and it drifts out, you cover the other eye and it drifts in.

Christopher Tyler has been studying traumatic brain injury. Strabismus may develop in patients with traumatic brain injury. The first thing that happens with a hit to the head is that your eyes converge, so traumatic brain injury has a direct connection with strabismus in the sense that it produces eye convergence. A model of football-related concussions showed that the forces to the brain predominantly affect the corpus callosum, the basal ganglia, and the upper brainstem. The latter two are the loci of control of vergence movements. A study of tissue shrinkage after traumatic brain injury had the same pattern of severity focused on the upper brainstem, which is where the binocular oculomotor control mechanisms are located. This seems to be a possible insight into why this sort of brain injury, perhaps twisting of the cranium on the spinal column, causes vergence problems.

In order to investigate these mechanisms in more detail, we developed a high-resolution fMRI prescription to look at the brainstem. Functional imaging of individual brainstem nuclei is possible.

In discussions with Dr. Rutar, we thought that intermittent exotropia was an interesting paradigm to investigate this kind of thing, because the patient can act as their own control. They can either be straight or take a strabismic angle. This allows you to do functional imaging comparison of the two states, back and forth, and look at the difference in activation. The subject is doing divergence movements for 12 seconds and then fixating for 12 seconds, and keeps alternating like that. We get a strong signal in the superior colliculus at the top of the brainstem, which is the main visual oculomotor control area, and a signal in the supraoculomotor area, which is the vergence control center. There is very little other activation anywhere in the brainstem. We possibly get a convincing signal at the level of the abducens nucleus bilaterally.

That is the kind of technique that we have developed that should be available for investigating the brainstem involvement in strabismus conditions.

K-Min Lee wondered if he looked at the pulvinar activity?

Christopher Tyler said that there was no other activation in the pulvinar.

K-Min Lee said that in order to interpret this result as a consequence of the vergence movement, rather than the motion signal in the visual input, you need to show that the activity from the superior colliculus is actually important, or is involved in driving this activity.

Cameron Parsa suggested that this fMRI technique could be useful for looking for the so-called disputed "divergence center" of the brain. Patients with Chiari malformation, where the cerebellum compresses the midbrain, have divergence insufficiency. Divergence insufficiency could result from an effect on the convergence center of the superior colliculus. If you had a patient with a Chiari malformation and you did this study, you might see the superior colliculus still lighting up or lighting up even more which would go along with the idea that the cerebellum compression is really inhibitory neurons to the superior colliculus. You could image a patient with Chiari malformation before and after suboccipital decompression.

Christopher Tyler said that this is exactly the kind of issue you can address with this fMRI technique.

Alan Scott wondered if you have tried to discriminate fusional vergence from typical motor near point convergence? The latter, of course, brings in the pupil and other things.

Christopher Tyler said that we'll put it on the slate.

OBLIQUE DYSFUNCTIONS

12. Clinical Problem: Under what circumstances should inferior oblique surgery be performed when inferior oblique overaction coexists with childhood strabismus?

Why this is a problem: Surgical correction increases anesthesia time, and adds some additional operative risks, whereas lack of surgical correction may require a second surgery in the future.

Michael Brodsky introduced the topic of oblique muscle dysfunctions by explaining the evolutionary origin of the oblique muscles. He referenced a *Nature* publication on why lizards have tails.

The oblique muscles are the lizard's tail of the eyes. They are these ocular appendages and what do they do? They are the only muscles that have contrived a functional origin near the front of the orbit and they "wag" near the back. Originally, in lateral-eyed animals, they produced torsion. But in humans, they prevent torsion. The pull of the rectus muscles would try to induce torsion. The obliques anchor the vertical retinal meridians in all directions so that you don't have torsion produced by the other extraocular muscles, and you don't have torsion produced by head or body inclination. The lizard's tail operates in the pitch plane - when the lizard pitches forward or back, the lizard's tail compensates. It is my belief that when you have POMA in strabismus, the disruption of binocular vision has changed your internal gyroscope so that your sense of vertical has actually tipped forward. So, the oblique muscles adjust. If the binocular equivalent of pitch is image tilt, so if you look at something that is slanted and you close one eye and then close the other eye, it will be tilted on each retina. I think that is what leads to torsional cyclodivergence that produces POMA.

Michael Clarke said that he doesn't personally have perfectly set criteria about when to weaken inferior oblique muscles in the presence of horizontal strabismus, particularly in a child who is not binocular. Sometimes, merely altering the horizontal alignment can affect the primary inferior oblique overaction that you see when the eyes are strabismic. Why do we weaken the inferior obliques? For purely aesthetic reasons? Or does doing so have some functional impact on the stability of results, even in a nonbinocular child?

Maria Arroyo Yllanes discussed clinical tests to determine what amount of hypertropia is due to inferior oblique overaction versus DVD in patients who have both coexisting. This is important in terms of surgical planning. She showed examples of patients with significant inferior oblique overaction and minimal DVD, who require only inferior oblique weakening, and patients with minimal inferior oblique overaction and significant DVD, who require DVD surgery. She described the Posner test for determining the presence of DVD.

Discussion:

Mauro Goldchmit said that in Brazil, they use measurements of hypertropia in lateral gaze as a way of distinguishing the relative contribution of the inferior oblique overaction compared to the DVD.

Maria Arroyo Yllanes said that she believes it is difficult to do these measurements, in part because DVD is so variable from hour to hour depending on fixation distance and lighting conditions.

David Guyton remarked that another way to tell is by looking at the fundus torsion. DVD does not show abnormal torsion of the fundus in straight-ahead gaze, with indirect ophthalmoscopy with the light turned down low. But you certainly see abnormal extorsion of the fundus in DVD coexisting with inferior oblique overaction. It doesn't help to distinguish how much of each might be present, but if you don't have any fundus extorsion, inferior oblique overaction is not contributing.

Another clinical difference is the speed of eye movement. You don't see this slow drifting movement with inferior oblique overaction. The eyes go right to their position.

Michael Graef said that the oldest test he knows to identify DVD is the dark red clusters of Bielschowsky.

Michael Brodsky mentioned that the one thing we found helpful is video oculography to distinguish how much is DVD and how much is true hypertropia. With video oculography, you can make it completely dark and assume that whatever hyperdeviation you see must be true hypertropia. Add on covering one eye to then see how much DVD is present. However, it is difficult to acquire the video oculography equipment.

He asked for clarification of the Posner test.

Maria Arroyo Yllanes answered that Posner said that the only way to equal the movements in DVD patients is to put both eyes in the same circumstances of light and fixation. So you put the occluder in front of one eye and that eye goes up. Then you put another occluder in front of the other eye and both eyes come down.

Arvind Chandna, also, mentioned measuring the hypertropia in lateral gaze: if it increases in abduction, DVD (which is created in part by action of the superior rectus) creates more hypertropia in abduction, whereas inferior oblique overaction creates more hypertropia in adduction.

Maria Arroyo Yllanes said that performing the Posner test is an alternative approach that does not require lateral gaze measurements.

Stephen Kraft stated that DVD could change. However, the hypotropia component that you measure in adduction is usually very stable. If you use a base-up prism in that position to neutralize the movement of the hypotropic eye, you then see the remainder of the DVD. Dr. Kraft also described the dynamics of the movements observed in the adducted eye under cover.

David Guyton said that another distinguishing feature is that a pure DVD rarely ever shows a V pattern, yet inferior oblique overactions almost always show a V pattern.

Richard Harrad concurred with Dr. Clarke, that he now rarely performs inferior oblique surgery at the same time as esotropia surgery. When the eyes are straight, the inferior oblique overaction is not usually visible and it also seems to diminish over a period of time.

Alan Scott, also, concurred. If you look at cases that are secondary exotropia that you reoperate on, they are almost all A patterns with superior oblique overactions. I never do inferior obliques unless the patient shows superior oblique underaction.

Scott Foster described his surgical algorithm. In large angle esotropias, it is difficult to see inferior oblique overaction or a V pattern. He performs a monocular recess resect procedure first, and then reexamines the patient. With the esotropia angle diminished, he can see the V pattern and the hypertropia in adduction, and then add on inferior oblique weakening (or superior oblique weakening if the pattern is the opposite). He also decides on whether to weaken the inferior oblique based on the degree of overaction.

Stephen Kraft said that with very large horizontal deviations, you cannot very easily tell whether you are seeing inferior oblique overaction, DVD, or both in adduction. By getting the eyes straighter, you can then assess the verticals more accurately, and make the best subsequent surgical decision.

13. Clinical Problem: Under what circumstances should inferior oblique surgery be performed when inferior oblique overaction coexists with childhood strabismus what causes A and V patterns? Could knowledge of the cause allow us to better predict who will develop A and V patterns, and how they should optimally be treated?

Why this is a problem: Patients who develop these patterns may require additional surgeries. Patients who fuse and have A or V patterns may adopt anomalous head postures.

André Roth wondered what is the appropriate surgical strategy for oblique overactions coexisting with horizontal deviations? He discussed the clinical assessments and surgical strategies for weakening the inferior oblique muscle. In the first approach, the vertical deviation is assessed in primary position and in lateral gaze of 25 degrees with the left eye and right eye fixing. Oblique muscle recession is indicated if there is a vertical deviation in the primary position and if it is greater in adduction than abduction. Dr. Roth recesses the inferior oblique muscle along its normal anatomical path. For the superior oblique, he cautions to not weaken it too aggressively; else binocularity in primary gaze will be affected.

Horizontal rectus muscle heterotopy can cause A and V patterns. Vertical shifts of the horizontal rectus muscle, or slanting their insertions, can efficiently collapse A and V patterns, but can increase cyclotorsion, which must subsequently be addressed by recession of the "overacting" oblique muscles.

Michael Brodsky mentioned heterotopic lateral rectus muscles. He wondered whether the origin is malpositioned, rather than the insertion, and that vertical offset of the insertion may actually exacerbate the abnormality in the muscle path.

André Roth said that I agree completely with you but it is not possible to change paths of the muscle and we can only change the insertion, and the effect of shifting the insertion is sufficient.

Omondi Nyong'o stated that the functional insertion or origin of the muscle might be at the muscle pulley.

Henry Metz discussed the importance of precise terminology. "Inferior oblique overaction" implies that the muscle is truly overacting. The observer can only be certain of the appearance of overaction in the field of the inferior oblique muscle (or superior oblique); the underlying etiology of this eye movement may or may not be due to the oblique muscle. For example, in a tight lateral rectus, the eye shoots up or down in adduction. And the same thing will happen in Duane syndrome. And we know if we loosen the tight lateral rectus or we do something about the co-contracting lateral rectus in the Duane syndrome, the appearance of the oblique overaction disappears.

In A and V patterns, why does a vertical shift of the horizontal rectus muscles eliminate or minimize the apparent oblique overaction?

Huibert Simonsz agreed that the terminology is flawed. For example, craniosynostosis syndrome with cyclorotation of the orbits does not have true "oblique overactions." In Germany, the confusing term "overaction" has been abolished in favor of purely descriptive terms like "upshoot in adduction," "downshoot in abduction" because those terms describe what actually happens.

Arthur Jampolsky agreed that appropriate terminology is imperative, and referenced an article by Walter B. Lancaster at the first New Orleans Symposium on terminology and oculomotility. Functional and descriptive terms should not be intermixed.

Harley Bicas stated that the superior rectus, not the inferior oblique, is the true elevator of the eye (even in adduction). Clinicians say that there is an overaction of the inferior oblique because they see the eye coming from one position and then suddenly elevating in the position where the inferior oblique acts more than in abduction.

David Romero Apis said the true elevator of the eye in adduction is the brain, which gives the order for the eye to elevate in adduction.

Huibert Simonsz invoked the pulley concept and its importance in understanding ocular rotations.

Harley Bicas said that in any ocular position, all six muscles combine to produce an eye movement and, of course, other structures, conjunctiva, Tenon's etc., are also involved. You may differentiate muscles by their many actions, but you have to compare the actions in different positions.

14. Clinical Problem: What causes congenital "superior oblique palsy," and should one tailor treatment to the underlying cause?

Why this is a problem: Not knowing the cause could lead to inappropriate/incorrect surgery.

David Guyton said that many cases of so-called "congenital superior oblique paresis" are really not paretic after all, but rather represent what we call basic cyclovertical deviations and fall into a previously unnamed group. In 1995, Joe Demer and his colleagues summarized their MRI findings of 19 superior oblique muscles, diagnosed to be palsies based on clinical criteria; MRI demonstrated that about half exhibit normal cross sectional size and contractile characteristics. Might these be simply the basic cyclovertical deviation, that we are now calling that, with no paresis after all?

Is there a normally occurring vergence that could cause a basic cyclovertical deviation by a muscle length adaptation if it was continually acting? A long chronic stimulation of a basic cyclovertical vergence could do that, if it was unchecked by fusional vergence. Chronic vergence stimulation can cause strabismus via a muscle length adaptation, if it is not counteracted by fusional vergence. The answer is "yes"; there is a mechanism. This is Jim Enright's mechanism from 1992. He took small prisms and built people up to fusing vertical prisms a small amount and then took the prism away, and he could tell with video oculography by looking at combined vertical and torsional movements, which extraocular muscles were primarily active. He found that most people do indeed use the oblique muscles to fuse small vertical deviations. There was some discussion here earlier what the oblique muscles were for. This is something they are used for in most of us right here. The right superior oblique, for example, would bring the right eye down and intort it, and the left inferior oblique muscle would bring this eye up and extort it. So you get a cycloversion combined with a vertical vergence. This mechanism is out there. It has been used to explain a whole lot of things since 1992. If low-grade stimulation from this particular mechanism continues long term, what will happen? The left eye will come up and will extort, the right eye will go down and intort. The result will be a misalignment, vertically and torsionally, that is very much the same and perhaps indistinguishable from a left superior oblique paresis. The left eye is up and extorted.

We built a tilting haploscope from an old arc perimeter, with video oculography cameras, to investigate what will happen when we create vertical misalignments in normal subjects. With this apparatus, we adapted the subjects to an increasing vertical disparity over thirty minutes. Using these concentric circle targets without torsional clues, the normal subjects would typically build up to three to five prism diopters of a vertical misalignment that persists for about two minutes once they are dissociated. After they are dissociated, we would tilt the whole apparatus forward, backwards, and forwards, and measure the misalignments with video oculography. We found that these vertical misalignments do change with head tilt to either side, like the head tilt changes with superior oblique paresis. So, this showed us that we could produce head tilt responsive vertical strabismus in normal subjects without any underlying cyclovertical paresis at all.

Using our tilting haploscope, we also recorded a number of patients with superior oblique paresis who can fuse. What we do is analyze these simultaneous vertical and torsional movements of the eyes during fusion, or the breaking of fusion, and from this we can identify which muscles are primarily being used for fusion. Interestingly enough, we found that most patients with superior oblique paresis use their vertical rectus muscles. Curiously enough, some use their oblique muscles, even though the superior oblique is presumably weak. And some use a superior oblique muscle in the affected eye and the superior rectus muscle in the other eye.

71

Now, we are finding that these three groups behave differently. First, regarding their response to the Bielschowsky head tilt test, and, secondly, their response to vertical rectus muscle surgery combined with inferior oblique surgery, with overcorrections common in one group but not in the other. So this is still an investigation very much in progress. We have to acquire much more data before making definitive statements. But we are getting at some of these different types of what we call "congenital superior oblique paresis" by the basis of how they fuse, which muscles they use, and maybe these studies will help us determine the best surgical approach in a given patient in the future.

Huibert Simonsz appreciated the term "basic cyclovertical deviations" because we need a term for nonparetic motor vertical disorders.

Cameron Parsa said that some of the very recent magnetic resonance (MR) data shows that there are smaller fourth nerves in the affected orbit. Perhaps if the nerves are sufficiently hypoplastic to fall beyond a certain threshold of necessary axons to fully innervate the superior oblique muscle, maybe those are the superior oblique muscles that are lax. Perhaps for those cases, superior oblique tucks would not be as useful as inferior oblique recessions. Maybe those are the cases where the vertical rectus muscles are being used more to fuse rather than the oblique muscles. In other words, when you find cases with the haploscope that oblique muscles are not being used to fuse, maybe those are the ones that there truly is some minimal threshold nerve paresis and lax tendons.

Christopher Tyler found it puzzling why vertical vergence is so much slower than horizontal vergence for fusion. Is that something that is understood?

David Guyton said that he didn't know. Panum's fusional area is very narrow vertically and maybe that is the reason.

Arthur Jampolsky asked what are the diagnostic criteria for a fresh and real superior oblique palsy in animals and humans? Those are different than the chronic ones for then you have other muscles that enter into the adaptation.

Cameron Parsa said that he thinks we have to distinguish between whether we are talking about congenital hypoplastic nerves vs. acquired atrophic ones.

David Guyton said we are called upon to make many larger and faster horizontal vergences than we ever are vertical; maybe that's why they never kept up.

Felisa Shokida described two cases of unilateral congenital superior oblique palsy with intriguing features.

  a. Why is the contralateral superior oblique muscle larger than normal in unilateral congenital superior oblique palsy?
  b. Why can a patient with a long-standing unilateral congenital superior oblique palsy develop an A pattern instead of the typically expected V pattern?

David Guyton said that clearly this is a true superior oblique paresis from the MRI findings, but without the V pattern. This is unusual. There must be something else going on here.

Michael Graef had a similar case recently of a congenital superior oblique palsy confirmed by MRI (showing a hypoplastic superior oblique), but with an A pattern and incyclotorsion.

David Guyton queried if she looked at the fundus torsion?

Felisa Shokida said that it appeared normal.

Scott Foster recently reviewed his series of congenital superior oblique palsies and surprisingly did not see large V patterns, only small V patterns or no V patterns.

Federico Vélez said that contralateral overdepression in adduction ("superior oblique overaction") is a very common finding in congenital superior oblique palsies. Joe Demer published about enlargement of the contralateral superior oblique.

Alan Scott stated that superior oblique palsy is unclear in its origins. It is very possible that these two superior oblique nuclei, which are very close to one another and actually go through one another, could develop asymmetrically. One side gets too few axons and the other side gets too many.

Michael Graef said that it might have something to do with compensation because he believes the inferior obliques are down regulated in these patients.

John Sloper wondered if you can explain any of these unusual findings (A pattern) due to an abnormal insertion of the superior oblique tendon, when it is too far back?

Tina Rutar was intrigued by Dr. Scott's comment. We are learning more and more that certain types of childhood strabismus are due to congenital cranial dysinnervation disorders. Could we use better imaging, or perhaps histology (after patients with congenital superior oblique palsy die) and investigate the fourth nuclei and fourth nerves better in these "congenital superior oblique palsy" patients?

Alan Scott said that, yes, and imaging, including functional MRI, continues to improve.

Huibert Simonsz stated that disruption of fusion allows basic cyclovertical disorders to become manifest. This can happen in healthy volunteers who are monocularly patched. Many will develop upshoots or downshoots in adduction. Also, Dr. Simonsz has detected robust contraction of the superior oblique muscle under succinylcholine anesthesia in a patient with clinical features of congenital superior oblique palsy, suggesting that the muscle need not actually be palsied to give the clinical picture of one. The Bielschowsky head tilt test can be positive due to a nonparetic motility disorder. He also discussed the importance of the vestibular system and cerebellum in adaption to vertical tropias.

15. Clinical Problem: What is the pathogenesis underlying congenital Brown's syndrome?

Why this is a problem: Incomplete understanding of the pathogenesis of Brown's syndrome may lead to inappropriate/ incorrect surgery.

Hermann Mühlendyck described his intraoperative findings in several cases of Brown's syndrome: exploration of the region of the superior oblique tendon toward the trochlea showed a fibrotic strand attaching near the trochlea to the orbital wall. The fibrotic strand may be an atavistic superior oblique. Excision of the insertional part of the fibrotic strand is an effective treatment.

Cameron Parsa said that he finds the fibrotic strands in Brown's syndrome to be very reminiscent of Duane syndrome. In Duane syndrome, the lateral rectus muscle, which isn't fully innervated even by the third cranial nerve, becomes fibrotic instead of lax. That was an enigma that Dr. Hoyt pointed out to me. Why would the muscle not be lax? Why is it fibrotic in Duane syndrome? And the answer might be that the orbital layer, which is supposed to differentiate into muscle, may not do so completely if it is not innervated. There are neurotrophic factors that help differentiate this layer into muscle. In the 1960s, Pabst and Stein showed EMG studies that there could be co-innervation of the superior oblique with the inferior oblique in Brown's syndrome and Orton, Bauer, and others have brought that up based on clinical findings. We see palpebral fissure widening on adduction and many of the clinical signs that would go along with a misinnervation syndrome. Just a few months ago, Francois-Xavier Borruat did some MR imaging of patients with Brown's syndrome and found that the fourth nerve was indeed very hypoplastic, much like we confirmed in Duane syndrome. Thus, Brown's syndrome could be a misinnervation syndrome with a fibrotic muscle and a secondary mechanical restriction created by the trochlea. In some cases, perhaps, Brown's syndrome could be relieved by surgery, although co-innervation would persist after surgery. There probably is a spectrum, just as we have different phenotypes for Duane syndrome.

Hermann Mühlendyck said that in these cases, the fibrotic strand was not contiguous with the rest of the superior oblique muscle: it attached to the orbital wall. A second argument against the misinnervation hypothesis is that Brown's syndrome patients have normal depression in adduction.

Another participant commented on "congenital" Brown's syndrome patients who appear to spontaneously improve, or later develop Brown's syndrome in the contralateral eye. These are young children with negative inflammatory work-ups.

Michael Graef mentioned that sometimes Brown's patients do show downshoots in adductions.

Keith McNeer said that the biggest distinction between primary Brown's syndrome and secondary Brown's syndrome is an overacting superior oblique, which implies something other than a fibrotic muscle. To make the distinction between congenital Brown's and any other kind of Brown's is very easy if you measure the depth perception in the field nasal to the adducted eye. If you have grown up with congenital Brown's, you don't have any depth perception there, whereas you have it everywhere else.

Dr. McNeer's experience in operating on Brown's syndrome has *not* been extremely rewarding; these patients are difficult to cure.

Arthur Jampolsky said that we have all cured Brown's syndrome intraoperatively and had free ocular rotations up and in, only to find Brown's syndrome recur in the postoperative period.

Hermann Mühlendyck said that trauma to the trochlear area can cause a hematoma and swelling in the area of the superior oblique tendon, and can create a transient Brown's syndrome.

Michael Graef reviewed the clinical features of Brown's syndrome: hypertropia increasing in upgaze, V or Y incomitance that we can explain as a bridle effect of the superior or inferior oblique muscles. If the hypotropia is in primary position, the patient adopts an anomalous head posture. There is widening of the palpebral fissure, which can also be explained as a bridle effect.

The etiology is heterogeneous and the underlying cause of Brown's syndrome can be located between the trochlea and insertion of the tendon, as described by Dr. Mühlendyck just in his talk, or within the trochlea itself, or with the tendon directly behind the trochlea. A congenital Brown's syndrome can also be caused by dysinnervation.

We perform a standard recession of the superior oblique tendon by 10 millimeters and only fix the anterior third of the tendon 12 millimeters away from the limbus. This is a reversible procedure. We do not use a tendon expander. We do not perform tenectomy. I have never done this and have never performed additional inferior oblique surgery for Brown's syndrome.

David Guyton asked whether Dr. Mühlendyck routinely exposed the trochlea when operating on Brown's syndrome patients, and wouldn't that create a lot of scarring?

Hermann Mühlendyck clarified that he carried the dissection of the fibrotic strand to the orbital wall in only a few patients. He initially thought that the entire fibrotic strand might need to be excised. However, he later modified the technique to merely disinsert the fibrotic strand from its insertion onto the globe.

16. Clinical Problem: How should one treat bilateral superior oblique palsy?

Why this is a problem: Various surgical options may not address both the excyclotorsion and the V-pattern esotropia, leaving patients diplopic in primary gaze and/or down gaze.

Anja Palmowski-Wolfe reviewed some of the surgical options for treating bilateral superior oblique palsy. With small vertical deviations in adduction, bilateral recessions of the inferior oblique muscles may be sufficient. An alternative is a superior oblique tuck. In larger deviations these may be combined. One can also tighten the superior oblique by pulling it forward somewhat like the Harada-Ito procedure, known as the adaptation of Fels. The Boergen procedure is popular; you create a Y split in the superior oblique tendon and use nonabsorbable suture to tighten the anterior part. The Boergen procedure has no risk of creating a Brown's syndrome and it works nicely for cases in which correcting excyclotorsion is most important. It is also reversible by removing the nonabsorbable suture. If there is more than 5 PD hypertropia in primary position, you can add to this procedure a tuck of the posterior fibers as well.

Carlos Souza-Dias described his surgical technique for treating bilateral superior oblique palsy without inferior oblique overaction. Superior oblique tucks are unsatisfactory, as they create an iatrogenic Brown's syndrome. His preferred technique is large bilateral inferior rectus recessions.

The adduction and extorsion forces of the inferior rectus are stronger than the abduction than the intorsion force of the superior oblique. If one weakens the inferior rectus (generally, Dr. Souza-Dias performs 6 mm recessions), the forces of the inferior rectus and the superior oblique are better balanced. He provided two clinical examples and mentioned that Burton Kushner had also previously presented a paper of six cases operated on with this technique.

Arthur Jampolsky said that Dr. Souza-Dias was absolutely correct in saying the problem was diplopia in downgaze, in reading position. Most people don't go far enough down to measure true reading position. We all know that in bilateral superior oblique palsies, from just above horizontal and upward, everything is usually fused fine. The patient habitually assumes a small chin down position to fuse and gets along very well with that. The real problem is in measurements made in the real-life downgaze reading position. Few manuscripts on the topic of post surgical measurement of bilateral superior oblique palsies provide real-life downgaze reading measurements.

Dr. Jampolsky lectured on this topic in Mexico: "If I had a bilateral superior oblique palsy, and I had a severe one with 25 eso downgaze and a lot of torsion, each eye, I wouldn't try to have it corrected. I would just tuck my head down a few degrees and be fine for everything. And I would put myself in monovision for near and you wouldn't touch me with a ten-foot surgical pole no matter how much I was promised by whatever surgeon." At the conclusion of that statement, Dr. Campomanes, Alejandra's mother, came rushing up from the audience and said, "Oh, Dr. Jampolsky, I have had a very successful career. I have a bilateral superior oblique palsy and no one has ever noticed it, nor have I ever had it fixed. I am a wonderful surgeon."

Dr. Jampolsky also commented to Dr. Simonsz, on the superior rectus overaction contracture syndrome. It brings into question the role of the otolith apparatus in driving a condition of a forced head tilt vertical difference (right and left head tilt). Superior rectus contracture and superior oblique palsy are the only two muscles that really cause a forced tilt difference.

Huibert Simonsz questioned Dr. Jampolsky on what is the evidence that the superior rectus is truly contractured. Could the clinical findings of "superior rectus contracture" be explained by other mechanisms, such as connective tissue changes or innervational changes?

Arthur Jampolsky answered that superior rectus contracture *is* a shortened superior rectus muscle as measured by length tension, by forced ductions, by clinical rotation observations and by forced head tilt vertical difference in the right and left tilt measurements. And - by predicting and observing the amelioration or elimination of the contracture - and its forced vertical tilt difference, simply by recessing the contracted superior rectus muscle, in itself, an amazing unpredictable, and unexplainable happenstance.

DISSOCIATED VERTICAL DEVIATION

17. Clinical Problem: Can we come up with a better solution for DVD other than surgically limiting ocular rotations?

Why this is a problem: DVD can recur unless you cripple the ability of the eye to move up, or establish sensory fusion. Limiting ocular rotations is not an ideal solution to the problem. Many patients with DVD do not have the ability to fuse.

Tina Rutar presented a question asked by Dr. Fredrick: if you have inferior oblique overaction and DVD coexisting, why is it so important to tease out the contribution of the inferior oblique overaction and the DVD? Why not do anterior transpositions of the inferior oblique muscle?

David Guyton answered that it causes a cosmetic deformity by raising the lower eyelid.

Federico Vélez said that he never places the inferior oblique anterior to inferior rectus. He does weaken the inferior oblique in cases where there is a DVD in the right eye looking left, and a DVD in the left eye looking to the right, and a small V pattern.

Bradley Black agreed with Dr. Vélez' approach, and also added that DVDs in such cases rarely measure 20 PD or more. His surgical technique is to suture that anterior corner of the inferior oblique immediately temporal to the inferior rectus insertion, and let the posterior corner of the inferior oblique drop back where it wants to.

Maria Arroyo Yllanes remarked that if the DVD is larger, you have to do something about it because if you weaken only inferior oblique by the traditional techniques, it isn't enough.

Mauro Goldchmit stated that the anterior transposition of the inferior oblique will correct 15 to 30 PD of DVD. It will bring the eye from this up position to the mid line. The surgeon must bring the anterior and posterior parts of the muscle together in one suture; this will not create any limitation of elevation. He occasionally performs inferior oblique anterior transposition even without inferior oblique overaction.

Scott Foster said that a manuscript on the treatment of DVD using various techniques recently appeared in IOVS. The best treatment is bilateral 10 mm SR recessions, regardless of the symmetry of the DVD.

Maria Arroyo Yllanes demonstrated videos of patients with DVD and clinical tests that bring out the DVD: a base down prism test, and the Posner test.

David Guyton described his surgical coil recordings of 10 patients with DVD, and provided clinical videos, in support of his hypothesis that DVD is a learned anticipatory response to help dampen latent nystagmus and thus improve vision.

Fixing with one eye when the other eye is occluded, or simply concentrating, seems to increase an imbalance of input to the vestibular system, and the result is latent nystagmus. Latent nystagmus appears horizontal and cyclovertical; cyclovertical latent nystagmus is similar to the vestibular nystagmus that normally occurs with dynamic head tilting. To improve vision, convergence occurs, the phenomenon behind DHD.

DHD was first described in the exotropic patient, where you let the good eye see. When you cover the good eye, the weaker eye has to see, and you get convergence. You see DHD with esotropic patients, with straight patients, too. The dissociated movement in DHD is a convergence. Also, a combination of vertical vergence cycloversions occurs, which we call DVD.

Can we treat the underlying cause rather than cripple the muscles, which is the topic of this session? In other words, can we decrease the intensity of the monocular fixation preference that brings out the latent nystagmus that causes this problem? For example, can we decrease suppression using some sort of biofeedback training? Can we harness the Bielschowsky phenomenon by disadvantaging the fixing eye to bring the other eye down? Can we use medications to dampen the latent nystagmus? Has anyone tried baclofen or neurontin to treat DVD? Or is there some surgical approach that we can take, not so much to cripple the muscles that are causing the DVD, but rather to decrease the slow phase of the latent nystagmus that is its underlying cause?

Tina Rutar asked Dr. Guyton for clarification re: harnessing the Bielschowsky phenomenon.

David Guyton answered that when you put a dark filter in front of the fixing eye, the other eye comes down. There might be some way to disadvantage the fixing eye enough to bring the other eye down without making the fixation switch to the other eye.

Tina Rutar wondered if anyone has tried Bangerter filters of various intensities to see if they can equalize the two eyes just enough so that DVD is not manifest in either eye?

Denise Satterfield said that it's important to distinguish whether the DVD is manifest or only present with occlusion. An occlusion hyperphoria doesn't require surgery.

K-Min Lee asked whether DVD patterns change with head positioning?

David Guyton said, yes, vestibular input from certain head tilts can somehow dampen latent nystagmus so that the patient does not need to use the DVD to dampen it

K-Min Lee mentioned a leaky neural integrator.

David Romero Apis showed an example of a case in which he successfully treated DVD with bilateral recessions of the inferior rectus muscles. The fixation duress that might drive DVD was corrected.

David Guyton remarked that one day postoperatively, the extraocular muscles are in a state of "shock" and that could explain the (temporary) disappearance of DVD. He operated on all four oblique muscles once, saw the DVD disappear on postoperative day one, only to have it recur six weeks later.

David Romero Apis said that the patient he treated with bilateral inferior rectus recessions came back for a four-month postoperative exam, and still had only a very small DVD.

David Guyton suggested that the reason Dr. Romero Apis was successful may be because weakening the inferior rectus muscles weakens/treats the slow phase of latent nystagmus directed downward.

NEW APPROACHES TO STRABISMUS MANAGEMENT

18. Clinical Problem: Strabismus management is essentially limited to refractive correction, amblyopia treatment, extraocular muscle injections, and surgery. How can we develop new solutions?

Why this is a problem: Other fields in ophthalmology have had treatment revolutions, but we are still employing many techniques developed decades ago.

John Brabyn said that in strabismus we often seem to be dealing with treating results rather than causes. In fact, much of the strabismus management that is currently used, was traditionally used and also new methods being worked on have a lot in common with rehabilitation. Even the same language can apply. For example, we use visual orthoptics, such as glasses and prisms. We use training methods such as an amblyopia therapy. And we cripple or strengthen the eye muscles by surgery or injections. Let's explore new approaches to strabismus management.

Harley Bicas discussed a novel concept/model on how to avoid undesirable post-operative drift.

We work with three basic types of forces in the oculomotor system. First come those originating from neural stimuli, which evoke muscular contractions, as that of the medial rectus to produce adduction, and which evoke muscular relaxation, as that of the lateral rectus to produce the same adduction. This is the so-called active force. When the eye moves, the conjunctiva, fascia, ligaments, and the muscles themselves are stretched so they evoke what we call the passive force. Active forces start a movement; passive forces stop it. But if this system were perfectly conservative, the eye would return to where it had departed, and the result would be a perpetual pendular-like movement. Fortunately, dissipative forces, that is, heat and friction, oppose the continuity of movement.

Our present surgical techniques attempt to weaken and to strengthen muscles and tissues, that is, to modify active and passive forces. When we cannot succeed in achieving stability of ocular alignment, perhaps we should attempt to modify the dissipative forces instead. For example, could we increase friction by injecting a viscous substance into the orbit? Could we increase friction by using magnetic fields? A small magnetic substance might be sufficient to stabilize the eye in a given position. To avoid a slow, large postoperative drift of the eye, one would require much less force, much less energy than what would be required to stop the fast phase of nystagmus.

Cameron Parsa discussed whether accommodation and the triad of accommodation/convergence/miosis may explain some features of sensory strabismus and the "esotonus" behind infantile esotropia.

Sensory strabismus was brought up in the oculomotor tonus symposium at Smith-Kettlewell six years ago. Children who lose vision in one eye tend to become esotropic, whereas adults tend to become exotropic. The idea that Worth brought up over a century ago that this was due to refractive error, that children were generally hyperopic and would become esotropic if they lost vision in one eye, and adults were generally myopic and, therefore, exotropic. That was contested by Gunter Von Noorden, who looked at some children with refractive errors and didn't find quite that correlation. If we look back at that paper, which is used to contest Worth's original observation, we see that refractive error wasn't included in the paper, and we don't know if the children had been corrected and observed or not. Why that matters is, of course, the issue of accommodation has to be addressed.

87

If we take our ability to modulate accommodation by optical correction, we can modulate the force generated by the medial rectus muscle as part of the near reflex triad to alter the position of the nonseeing eye in the orbit. So, we have a hyperopic child with a blind eye, we can leave the correction out, for example. They will be accommodating more and that will drive the eye in. If we see the eye is turning too far in, and we want to avoid that esotropia, we will give them their correction. If they are still esotropic, give them bifocals, and it works. This is what I do routinely in the clinic or what I teach our fellows and residents. By the time you see the patient they are already esotropic and secondary medial rectus muscle contracture has occurred. Giving them bifocal glasses may reverse part of it, but not all of it. But then we use Botox to straighten the eyes and then put them in the bifocal glasses. If it is an adult and they are already exotropic because they have been myopic and haven't been accommodating as much, Botox to the lateral rectus muscles may not work as well. But if we use bupivacaine to the medial rectus muscle and Botox to the lateral rectus muscle, we might get an alignment, which, again, we then modulate with the use of glasses. We can use overminus correction to elicit accommodation in those adults that can do it and keep the eyes straighter. This is a rather special situation in sensory exotropia, but it brings to the fore the idea that the near reflex triad is an essential component in strabismus, even when there's no binocular potential.

So, we don't have binocular potential in this one eye that's blind, and we also don't have binocular potential in the early months of life before those connections have been made in the cortex. So, if we see a child with esotropia in the first few months of life that may be a place where we really should intervene aggressively with refractive correction, either glasses or contact lenses, to avoid that contracture of the medial rectus muscle, which would otherwise occur because of the near reflex triad. Once the fusional potential does exist in the cortex, those connections are made; the eyes are relatively straight and then can lock in on target.

In the studies that have looked at optokinetic nystagmus (OKN) nasal temporal disparity and pursuit disparities with the nasal toward bias, the issue of the near reflex triad has been overlooked. The near correction was not given to those subjects. So, in order to elicit optimal acuity, they would accommodate, eliciting convergence, which would create this medial rectus muscle esotonus.

André Roth said that conventional surgery is a mechanical adjustment of muscle forces in the context of a physiological system. The results are predictable within the limit of probability, with a mean effect and associated variance.

Passive muscle forces may be assessed under general anesthesia. The basic angle is a minimum angle in esotropia and the maximum in exotropia. Eye position is observed under anesthesia with muscle paralysis. In this example you see a child with an infantile esotropia with a basic angle of five degrees and an asymmetrical position. The right eye is slightly divergent and the left eye visibly convergent.

Then we measure muscle extensibility and calculate the extensibility differential, which is the extensibility of the lateral rectus minus the extensibility of the medial rectus. And this extensibility is measured with a spring.

Such parameters should be taken into account when determining surgical dose, and we use this method systematically to improve the predictability of conventional surgery. Dr. Roth showed a surgical table that is found in his strabismus textbook and can be used to adjust conventional surgical doses by 0.5 – 2.0 mm.

Michael Graef asked whether Dr. Roth measures axial length and takes that into account?

André Roth said that we should take dimensions of the eye into account. He reduces surgical dose in patients with hyperopia greater than four diopters.

Mauro Goldchmit agreed with Dr. Roth that intraoperative measurements are important in modulating surgical dose. During his early career, Dr. Goldchmit used to plan the surgery in the office and then go to the operating room and do exactly what was in the chart. By the time he started paying attention to eye position under anesthesia, forced duction tests, the spring-back test, etc., his results improved significantly.

David Guyton asked what is muscle length adaptation? He said hI was introduced to the concept by a transcript of the proceedings of a Smith-Kettlewell conference held about 20 years ago. You won't find this term in most strabismus texts. But if you consider that the half-life of the contractile proteins in adult skeletal muscles is only seven to 15 days, it is not surprising that the muscle physiologist in France and England discovered in the 1970s and 80s that skeletal muscles intrinsically adapt their lengths for optimal function and tension over the usual range of motion. They do this by adding or subtracting sarcomeres in response to the stimuli. Alan Scott showed that this also happens in extraocular muscles.

What are the primary stimuli involved in muscle length adaptation? One is tension, another is mechanical stretch, and the most important is neurologic stimulation. In particular, increased stimulation shortens muscles by a loss of sarcomeres and decreased stimulation lengthens muscles by a gain of sarcomeres. And this role that stimulation can play in muscle length adaptation suddenly completes a dynamic feedback system for maintenance of long-term ocular alignment. Retinal image disparity causes double vision, which the brain doesn't like. This leads to fast fusional vergence. Fast fusional vergence is the primary stimulus for vergence adaptation, which is a neurological means for fine-tuning functional muscle length over time. Both fast fusional vergence and vergence adaptation supply the necessary sufficient stimulation to drive muscle length adaptation in the proper direction for long-term maintenance of ocular alignment, despite growth spurts and changing ocular conditions. Note that each of these mechanisms acts to reduce the retinal image disparity and the original double vision. Also, most important, each mechanism acts to reduce the strain on the mechanism that precedes it. In these schemes, extraocular muscles do not adapt the position to which they are held but rather do whatever vergence stimulation they are subjected to.

How can we apply this knowledge to the prevention and treatment of strabismus? We should treat strictly accommodative esotropia early and fully with glasses because it's the excess convergence that shortens the medial rectus muscles, over time, resulting in a basic component to esotropia that requires surgery. We should strongly advocate bifocals or reading glasses for early presbyopes if they have poor binocular function because their increasing effort to accommodate is accompanied by increased convergence tonus, shortening the medial rectus muscles over time by inappropriate muscle length adaptation resulting in the reoccurrence of esotropia that many of them had as children. There is no question that we should postpone the decision on which eye or eyes to operate on, especially with reoperations, until we can observe eye position under anesthesia. Drs. Roth and Jampolsky have been advocating this for years.

Next, with the understanding that it is primarily the vergence stimulation that regulates extraocular muscle lengths over time, one is tempted to advocate various types of selective stimulation of vergence, such as fusional vergence exercises to effect changes in ocular alignment over time. Also, version stimulation such as range of motion exercises wouldn't be expected to alter alignment over time at all. Perhaps in the future we will work out ways to correct strabismus by mild external electrical stimulation of vergence, selectively. Perhaps, when sleeping.

91

John Sloper remarked that if this is the case why do we get intermittent exotropia?

David Guyton said that the mechanism that is causing intermittent exotropia is overcoming these factors.

Discussion:
Several discussants commented on the position of eyes while asleep and under deep anesthesia.

Arthur Jampolsky said that under anesthesia and in sleep, intermittent exotropes have eyes in the exotropic position. You don't have vergences at night unless you have bad dreams, but the continued or prolonged position of the eye also allows shortening of the muscle. So, if you did a big recession of a lateral rectus muscle on whatever patient, it gets shorter ("takes up the slack") by prolonged positioning.

Federico Vélez described functional electrical stimulation in patients with cut laryngeal nerves. It is possible to use diaphragm stimulation to stimulate their vocal cords, allowing recovery of voice. Dr. Vélez showed how open loop functional electrical stimulation could also be used to recover contraction of palsied extraocular muscles, controlling the movement of one eye with the contralateral eye according to Herring's law. The late Art Rosenbaum was instrumental in developing this line of research.

In Dr. Vélez' experiments, implanted electrodes in the medial rectus muscle of one eye were connected with the lateral rectus of the other eye in a model animal. A coil system identified the position of the eye in all gaze positions. Dr. Vélez showed that he could contract the lateral rectus with stimulation to the contralateral medial rectus.

Alan Scott remarked that he has been developing a similar concept in an animal model of blepharospasm. A small amount of neural stimulation (compared to the larger amount needed for direct muscle stimulation) contracts the levator palpebrae muscle, allowing the eye to open. In his research set up, he can stimulate the lid to come up and relax stimulation to allow the lid to come down. Developing long-term stable electrodes will be necessary to make this practical to patients.

Scott Foster said that the best springs that we have right now are the extraocular muscles. They are tonic springs. Transposition procedures work adequately in the surgical treatment of extraocular muscle palsy. Physical springs that require implantation into the orbit don't work, as they scar down. The future will bring nano machines: small molecules, which act like a machine. They produce ATP, which then goes through hydrolysis to produce mechanical energy. The two best examples are myosin, which we all know about, and kinesin, which is used for tubular transport, and these will probably be the new thing for muscle replacement. David Stretavan at UCSF is exceptionally knowledgeable in this subject.

Tina Rutar concluded that in the field of retina and other ophthalmological subspecialties, implantable micro sustained-drug delivery devices are very popular. These could be applied to strabismus. We use botulinum toxin frequently, but we do episodic injections into muscles. What if we had a slow release Botox capsule in the extraocular muscle that we were able to control, externally from the patient, and modulate the amount of Botox that was released over time, depending on the patient's eye alignment? And other pharmacological agents (bupivacaine?) could be slowly released, via external control, to strengthen eye muscles.

Index

A

A and V patterns, 8, 9, 67, 68

abduction, 14, 65, 67, 68, 69, 79

accommodation, iii, 2, 3, 19, 20, 23, 24, 25, 26, 27, 28, 48, 52, 59, 87, 88, 89, 91

active force, 86

adduction, 14, 20, 48, 65, 66, 67, 68, 69, 74, 76, 77, 79, 86

adjustable sutures, 53

alignment, 1, 14, 17, 50, 52, 54, 64, 87, 88, 90, 91, 93

alternate, 15, 16, 17, 47, 55, 56, 57, 61

amblyopia, iii, 4, 5, 6, 7, 11, 16, 30, 31, 32, 34, 35, 36, 37, 38, 39, 41, 44, 47, 86

amblyopia treatment, iii, 11, 16, 31, 34, 35, 86

ametropia, 17

aneisokonia, 44

anesthesia, 1, 8, 13, 15, 18, 19, 53, 63, 75, 89, 90, 91, 92

anisometropia, 5, 32, 34, 39, 44

anomalous pathways, 15

anomalous retinal correspondence, 53, 57

A pattern, 9, 66, 73, 74

arc of contact, 3

Arroyo Yllanes, Maria, xiv, 8, 10, 64, 65, 82, 83

ATP, 93

atropine, 25, 26, 27

axial length, 26, 89

B

Baggolini, 40

Bangerter, 45, 84

base out prism, 22, 27

base-up prism, 66

Bicas, Harley E.A., vii, 2, 11, 26, 32, 69, 86

Bielschowsky, 40, 65, 72, 75, 84

bifocals, 3, 28, 29, 46, 88, 91

binocularity, iii, 4, 5, 6, 13, 14, 15, 16, 17, 19, 21, 25, 29, 31, 36, 37, 39, 40, 41, 42, 43, 44, 45, 47, 57, 60, 63, 64, 67, 88, 91

biofeedback, 84

Black, Bradley C., vii, 3, 4, 23, 24, 26, 43, 82

blindness, 37, 88

Boergen, 79

Borruat, Francois-Xavier, 76

Botox, 17, 18, 41, 42, 51, 88, 93

botulinum toxin, also see Botox, 1, 13, 15, 93

Brabyn, John A., v, vi, vii, 10, 86

occlusion, 4, 6, 15, 16, 32, 34, 44, 45, 84
occlusion nystagmus, 59
ocular alignment, 1, 14, 87, 90, 91
ocular dominance, 13, 56
oculography, 14, 58, 65, 70, 71
oculomotor, 60, 61, 86, 87
Odom, J. Vernon, xii, 4, 30, 32
OKN, see optokinetic nystagmus, 89
onset, 1, 13, 14, 25, 40
ophthalmoscopy, 32, 64
optic neuritis, 36
optokinetic nystagmus, 89
optotype, 30, 31, 32, 33
orthoptics, 36, 86
orthoptist, 28, 36, 43, 52
orthotropia, 14, 16, 17, 42, 47, 52, 54
overcorrection, 7, 29, 47, 49, 52, 53, 54, 72
overminus, 49, 52, 88

P

Palmowski-Wolfe, Anja, v, xii, 4, 5, 6, 7, 8, 10, 35, 47, 79
palpebral fissure, 76, 78
Panum's, 17, 45, 72
paresis, 3, 70, 71, 72, 73
Parkinsonism, 34
Parsa, Cameron F., xii, 2, 5, 7, 9, 11, 25, 27, 34, 45, 53, 61, 72, 73, 76, 87
parvocellular, 36

passive forces, 86, 87, 89
patching, 16, 31, 32, 35, 36, 48, 49
Pediatric Eye Disease Investigator Group, 18, 49
penalization, 4, 34
peripheral, 40, 45, 56, 57, 58
phospholine iodide, 3, 28
photophobia, 58
plasticity, iii, 4, 5, 34, 35, 36, 37
Polat, Uri, 35
POMA, 19, 21, 63
Posner, 64, 65, 83
postoperative diplopia, 6, 41, 43, 44
postoperative drift, 11, 87
Pratt-Johnson, John, 50
preoperative, 7, 17, 29, 47, 48, 50
Prieto-Díaz, Julio, 15
primary oblique muscle overaction, also see POMA, 19
prism, 7, 17, 21, 22, 27, 29, 41, 43, 45, 47, 48, 49, 66, 70, 71, 83, 86
base out prism, 22, 27
base-up prism, 66
ptosis, 1, 13
pulley, 28, 68, 69
pulvinar, 61
pupil, 59, 62

R

recession, 4, 28, 29, 49, 53, 54, 67, 78, 92
rectus muscle, 2, 15, 18, 51,

Addendum

FRONT ROW, L TO R: **Tony Norcia, Tina Rutar, Michael Baiad, Anja Palmowski-Wolfe, John Brabyn, Art Jampolsky, Arvind Chandna, Lora Likova, Cameron Parsa, Renuka Rajagopal**

SECOND ROW, L TO R: **Alberto Ciancia, Keith McNeer, Henry Metz, Herb Simonsz, André Roth, Felisa Shokida, Mauro Goldchmit, Hermann Mühlendyck, Harley Bicas, Michael Graef**

THIRD ROW, L TO R: **Omondi Nyong'o, David Romero-Apis, K-Min Lee, Carlos Souza-Dias, Maria Arroyo-Yllanes, Denise Satterfield, Brad Black, Vernon Odom, Michael Brodsky, Alejandra de Alba Campomanes**

BACK ROW, L TO R: **Bob Johnson, Scott Foster, David Guyton, Jonathan Horton, John Sloper, Michael Clark, Christopher Tyler, Federico Vélez, Richard Harrad, Stephen Kraft**

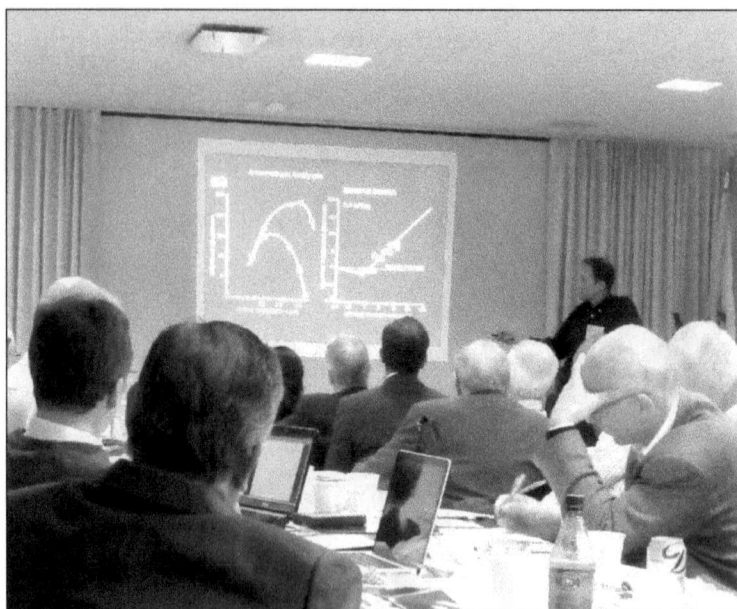

113